MODELLING PROBLEMS
IN CRACK TIP MECHANICS

TENTH CANADIAN FRACTURE CONFERENCE, CFC10

University of Waterloo
Waterloo, Ontario, Canada
August 24–26, 1983

Honorary Chairman

David M.R. Taplin *University of Dublin*

Organizing Committee

Chairman: Jerzy T. Pindera *University of Waterloo*
Vice-Chairmen: Donald Mills *Ontario Hydro*
               James W. Provan *McGill University*
Secretary: Bogdan R. Krasnowski *University of Waterloo*

Advisory Committee

J.J. Kacprzynski *National Research Council, Ottawa*
A.S. Kobayashi *University of Washington*
H.H.E. Leipholz *University of Waterloo, CSME*
R.J. Pick *University of Waterloo*
J.W. Provan *McGill University*
G.C. Sih *Lehigh University*
C.W. Smith *Virginia Polytechnic Institute and State University*
W.R. Tyson *Department of Energy, Mines and Resources, NRC*
M.P. Wnuk *University of Wisconsin – Milwaukee*

Cooperating and Sponsoring Organizations

The Canadian Committee for Research on the Strength and Fracture of
  Materials
University of Waterloo
Canadian Society for Mechanical Engineering, CSME
Natural Sciences and Engineering Research Council of Canada
Institute for Experimental Mechanics – IEM, University of Waterloo

Exhibitor

Intertechnology Inc., Toronto

Modelling Problems
in Crack Tip Mechanics

Proceedings of the Tenth Canadian Fracture Conference, held at the University of Waterloo, Waterloo, Ontario, Canada, August 24–26, 1983

JERZY T. PINDERA (editor)
Institute for Experimental Mechanics-IEM
University of Waterloo
Waterloo, Ontario, Canada

BOGDAN R. KRASNOWSKI (technical editor)
Institute for Experimental Mechanics-IEM
University of Waterloo
Waterloo, Ontario, Canada

1984 **MARTINUS NIJHOFF PUBLISHERS**
a member of the KLUWER ACADEMIC PUBLISHERS GROUP
DORDRECHT / BOSTON / LANCASTER

Distributors

for the United States and Canada: Kluwer Academic Publishers, 190 Old Derby Street, Hingham, MA 02043, USA
for the UK and Ireland: Kluwer Academic Publishers, MTP Press Limited, Falcon House, Queen Square, Lancaster LA1 1RN, England
for all other countries: Kluwer Academic Publishers Group, Distribution Center, P.O. Box 322, 3300 AH Dordrecht, The Netherlands

Library of Congress Cataloging in Publication Data

```
Canadian Fracture Conference (10th : 1983 : Ontario,
   Canada)
   Modelling problems in crack tip mechanics.

   Includes index.
   1. Fracture mechanics--Mathematical models--
Congresses.  I. Pindera, Jerzy-Tadeusz.  II. Krasnowski,
Bogdan Roman.  III. Title.
TA409.C36  1983          620.1'126          84-16703
ISBN 90-247-3067-8
```


ISBN 90-247-3067-8

PRINTED IN THE NETHERLANDS

Contents

Foreword

The general objective of the Tenth Canadian Fracture Conference was to
respond to progress in the engineering sciences - in particular with respect to rapidly developing new trends in the theory and methodology of research
and designing - and to the resulting needs of practical engineering in the
specific field of fracture mechanics and related areas of engineering
mechanics.

The basic underlying issue is the theory and practice of
physical analytical and iconic (reduced) modelling of the actually involved
physical processes and of the responses of physical bodies and systems to
actual energy flow - a problem which is becoming dominant in all fields of the
natural sciences. Accordingly, the theme of the CFC10 was "Modelling
Problems in Crack Tip Mechanics", a well defined and limited subject, the
scope of treatment of which can be as deep and as comprehensive as an involved researcher wishes it to be.

However, it must be recognized that the engineering sciences
belong to the social sciences too, and that therefore the cost effectiveness
in research and designing is one of the major engineering criteria. This
recognition influenced the policy of selecting papers for presentation and
publication. As a result, the majority of the presented papers published
the Proceedings is within the mainstream of the Conference and stresses
the problem of modelling of events of interest in crack tip mechanics, while
some papers present results of research conducted along lines traditional
in general engineering, to provide a possibly representative picture of
the field. Thus the Proceedings contain material which can be used to
analyze the advantages and disadvantages - from both the scientific and
utilitarian points of view - of both approaches in Fracture Mechanics:
the traditional approach based on a set of convenient, mostly intuitive,
assumptions, and the new approach developing along theoretical physical
lines in accordance with Bohr's correspondence principle. I expect that
this structure of the Conference and the Proceedings will give the reader
a certain perspective, will enable him to assess critically the developing
trends and new ideas in research in Fracture Mechanics, and to assess the
ranges of applicability of traditional approaches and of traditional

underlying intuitive assumptions. In other words, a kind of modernized Hegelian approach has been chosen in designing the program of the CFC10 and the format of the Conference Proceedings, to provide a stimulating framework for the exchange of ideas, conceptions and information.

Nevertheless it has been accepted, as an axiom, that the new high technology requires more advanced and more comprehensive theoretical bases of experimental methods in Fracture Mechanics with regard to determination of stress/strain/deformation/defect states and material responses than those presently prevailing in general engineering research and development.

Following the above presented general framework, an effort has been made to bring together individuals who are leading in basic and applied research and who are developing new concepts and models of reality directly or indirectly applicable to the mechanics of regions of cracks and notches, and individuals who are engaged in more conventional industrial research.

Within the general topic of Modelling Problems in Crack Tip Mechanics, several particular issues appear dominant at the present time. They deserve to be identified, and therefore are mentioned below.

Methods of fracture mechanics are based mostly on the criterion of strain energy density using various models of energy balance. For a long time the states of stress and strain in the regions of cracks and notches had been essentially analyzed as two-dimensional, linear elastic problems. Only recently, during the last decade, attempts have been made to develop reliable mathematical models and related analytical solutions for the actual three-dimensional stress-strain states in the regions of cracks and notches. During that period, analytical and numerical modelling of stress states in crack regions has been extended outside of the framework of the linear elastic model to consider the actual elasto-plastic or viscoelastic behaviour of ductile materials.

However, in experimental simulation of the physically nonlinear material behaviour in crack tip regions, using actual engineering materials, the criteria for determination of the boundary between the linear and nonlinear zones are usually not defined unequivocally. This fact makes it difficult or impossible to compare results of various experimental studies of the size and shape of plastic zones in the regions of crack tips. There a need to further develop the theoretical basis of experimental techniques for determination of plastic zones.

The problem of geometric nonlinearity is already within the range of interest of fracture mechanics since it is known that the very high stress values at the crack tips together with the high strain gradients produce geometric nonlinearity, the effects of which cannot be described using the concepts of classical elasticity theory.

Experience shows that the need exists to discuss more extensively such concepts as the plane stress state, plane strain state, generalized plane strain state, etc., and their compatibility with the basic assumptions of solid body mechanics.

It is also known that presently used experimental methods of analysis of three-dimensional stress states in crack tip regions are based on various assumptions, the reliability of which is seldom proven. Obviously, a need exists for more reliable and efficient methods of three-dimensional experimental stress analysis.

There is a noticeable lack of reliable mathematical models relating the actual responses of the microstructure of engineering materials to predictions of the very simplified mathematical models of materials designed as isothermal continuous bodies. In particular, there is no evidence that the thermoelastic effects occurring at grain boundaries can be neglected, and the transition from the actual dynamic fracture process to the quasi-static model of crack propagation is seldom analyzed.

The very particular problem of the existence of singularities in solutions for stress/strain states is still approached reluctantly, despite the fact that - in the opinion of many theoreticians - removal of singularities from solutions is one of the main theoretical problems in physics.

Summarizing, the Tenth Canadian Fracture Conference was organized and the Conference Proceedings are published to provide a forum for dissemination of accumulated knowledge on the actual strain and stress states in crack and notch regions, on the influence of the simplifying assumptions taken as the basis of various physical and mathematical models and resulting analytical relationships, and on the resulting reliability of the developed analytical and experimental methods and techniques, many of which simulate actual events using reduced (iconic) models. It is expected that the resulting discussions will lead to the development of more comprehensive and noncontradictory formulations of these theoretically and technologically important problems. It must be clearly stated here that the time when the accuracy and resolution of measurement systems represented major limiting factors in experimental research is in the past; at the present time the major limiting factors in understanding and predicting the behaviour of elastic and inelastic bodies needed for safe strength optimization are related to the lack of sufficiently comprehensive physical and mathematical models of the involved phenomena.

The organization of the CFC10 coincided with the organization of the Institute for Experimental Mechanics - IEM, University of Waterloo; the topic of CFC10 and the chosen approach are within the program of activities of the Institute, and thus the Conference became the first activity of the IEM in the field of engineering sciences.

It is my pleasant duty, as chairman of the CFC10 and editor of the Conference Proceedings, to acknowledge the fact that the Conference and the Proceedings are the results of the efficient and dedicated co-operation of a large group of persons and institutions, and to express my sincere appreciation of that co-operation and of the given support.

The concept of the CFC10 developed during a discussion of crack tip problems with David M.R. Taplin, former chairman of the Canadian Committee for Research on the Strength and Fracture of Materials.

The support given by the Natural Sciences and Engineering Research Council of Canada made it possible to assure the desired level of the Conference.

The encouragement and support given by the academic executives of the University of Waterloo, in particular, T.A. Brzustowski, Vice-President Academic, W.C. Lennox, Dean of Engineering and R.C.G. Haas, Chairman, Department of Civil Engineering, and their personal involvement, assured a proper format for the Conference.

The extent and good spirit of co-operation with the Conference Co-Chairmen, Donald Mills and James W. Provan, made my task much easier.

Very valuable suggestions and comments were given by several members of the Advisory Committee which influenced the scope and depth of treatment of the discussed topics.

I am very obliged to all the Conference Authors and participants for their willingness to share their ideas and research results by

presenting papers and contributing extensively to the discussions, particularly during the closing sessions.

In particular, I am deeply obliged to all the invited speakers
for their conceptual and factual contributions in the form of General
Lectures and Particular Basic Problems Lectures; their invaluable contributions have determined the profile of the CFC10 and of the Proceedings.

I am thankful to the Session Chairmen for accepting their difficult duties and for efficient chairing of the sessions.

Close co-operation with the Canadian Society for Mechanical
Engineering, in particular with F.P.J. Rimrott and H.H.E. Leipholz, is
gratefully acknowledged.

Intertechnology Ltd., Canadian manufacturer and distributor of
laboratory equipment and a co-sponsor, in the persons of Hans J. Jochem,
President, and his staff, demonstrated measurement systems being used in
modern engineering research, contributing to the necessary link between
theory and practice in research.

I also wish to express my sincere appreciation to D.E. Grierson
and his staff in the Solid Mechanics Division, University of Waterloo,
for their efficient and elegant editorial work in preparing these Proceedings.

My particular thanks are due to my close co-worker Bogdan Roman
Krasnowski, who accepted the difficult tasks of Secretary of the Conference and Technical Editor of the Proceedings, and performed them
well beyond the call of duty.

Jerzy-Tadeusz Pindera
Editor
Chairman, CFC10
Director, Institute for
Experimental Mechanics
University of Waterloo
March, 1984

Preface of the Honorary Chairman

This is the tenth in a series of national fracture conferences organized by the Canadian Committee for Research on the Strength and Fracture of Materials. This Committee was originally established ten years ago with the purpose of organizing the Fourth International Fracture Conference (ICF4) which was held with considerable success under the auspices of the Faculty of Engineering, University of Waterloo, in 1977. Since this time, national fracture conferences have been held regularly across the country. The orientation of all these conferences is necessarily *applied* and the ambience *informal* with much conviviality and "back of an envelope" discussion. It is a special pleasure that the tenth conference returned to Waterloo. Professor Pindera and Dr. Krasnowski have done a superb job of assembling an excellent company of scientists and engineers and in editing a lasting volume of Proceedings.

After ten years as Chairman of CSFM and fifteen years in the Faculty of Engineering at the University of Waterloo, I am now moving on to the green pastures of Trinity, Ireland. Already, the First Irish Durability and Fracture Conference (IDFCI) has been held and the proceedings published, and I look forward to close co-operation on fracture research between CSFM and IDFC, as well as between Trinity College and the University of Waterloo.

There are now some 40 countries which participate in the International Congress of Fracture via *National* committees of the CSFM mode. These countries variously organize national fracture conferences. As well there are *Regional* groups of fracture under the overall ICF umbrella. The European Group on Fracture (EGF) is the oldest and the next EGF conference is in Lisbon, Portugal, September 16-24, 1984. The Asian and Australian Fracture Group (AGF) has organized AGF-ICF1 in Beijing for November 22-26, 1983 and the Fracture Group of the Americas (FGA) has organized a Regional ICF Conference for November 28 - December 1, 1983 in Rio de Janiero, Brazil. The next quadrennial "olympus" is the major Sixth International Conference on Fracture, ICF6 which is being held in New Delhi, India, December 4-10, 1984. On behalf of ICF I welcome attendance at all these various ICF events.

The Canadian Committee has been a particularly active National Arm of ICF. This latest conference has contributed significantly to furthering the objective of fracture control and prevention and to fulfilling the wider objectives of the world organization. The Faculty of Engineering of the University of Waterloo can be justly proud of its contribution at CFC10.

D.M.R. Taplin
Honorary Chairman, CFC10
Trinity College
Dublin, Ireland

Preface of the Vice-President Academic

It is a pleasure to introduce the Proceedings of the Tenth Canadian Fracture Conference at Waterloo. When I first heard from Professor J.T. Pindera about a proposed conference on "Modelling Problems in Crack Tip Mechanics", I was struck by the timeliness of the theme. CFC10 would be yet another manifestation of what I perceive as a major thrust in modern mechanical engineering, namely using the microscopic to achieve the macroscopic.

Examples abound. The pervasive applications of microelectronics depend on the reliable performance of mass-produced integrated circuits whose elements are measured in microns. Microfabrication techniques are being used to produce miniature electromechanical devices, such as accelerometers. The technique of the ink-jet printer has its roots in the Millikan oil-drop experiment to measure the charge of an electron. The performance of disk memories for large computers is affected by the microscopic aerodynamics of the very thin air layer between the magnetic head and the surface of the rotating disk. Microspheres are used to provide thermal insulation, and an even newer development, superinsulation, involves plastic sheets coated with metal films so thin that the heat transfer through them cannot be described in terms of bulk properties. Even such an old problem as the heat transfer across a lightly loaded contact between a plane and a cylinder has a modern microscopic dimension. To calculate the heat transfer rate accurately, it is necessary to include the molecular flow of gases in that part of the gap near the contact where the dimensions are too small for the gas to be treated as a continuum.

It seems that similar considerations may have relevance to crack propagation, a process which I construe as connecting the microscopic and macroscopic manifestations of failure. They suggest some challenging questions. For example, microfabrication is emerging as a scientific tool which can help define the limits of continuum behaviour in solids. Could it be used in experiments on crack-tip behaviour in which the microscopic influences are varied in a controlled fashion? It is becoming possible to make measurements which would define the limits of applicability of the continuum models? Is it now becoming possible learn enough of the physics of the microscopic processes involved in the propagation of cracks to produce scaling laws which account for all the important influences?

Some such questions are being answered now. Others cannot be answered yet, but are being addressed in current research. Still others will be the subject of future work. CFC10 and future conferences in this series will undoubtedly provide the forum where they can be discussed.

T.A. Brzustowski
University of Waterloo

Welcoming Address

The invitation to present some brief comments to this conference gives
me the opportunity to welcome participants in my role as Chairman of the
Department of Civil Engineering, to provide a few observations on an area
of fracture which I deal with and finally to offer a couple of suggestions
which you may consider as you see fit.

This conference has delegates from a wide range of geographical
locations and a wide range of institutions, government and industry. They
all have considerable background and knowledge in the area and many have
particularly eminent reputations. It is rare and fortunate when so much
expertise can be brought into a single conference of this type. Consequently, it gives me particular pleasure that the Conference was held here
at Waterloo and to welcome all the delegates on behalf of my Department.

It also gives me considerable pleasure to congratulate Professor
Jerzy Pindera, Dr. Bogdan Krasnowski and their colleagues for such an
excellent job in organizing the conference. It is particularly fitting
that this conference is so well organized and of such obvious high quality
because it provides a sort of "launching pad" for the new Institute for
Experimental Mechanics. Congratulations should again go to Professor
Pindera who has been named the first director of the Institute.

Now I would like to comment on an area of fracture that I work
in which does not have the sophistication and depth of knowledge represented in the crack tip modelling problems described in this conference.
But, what it lacks in this way is more than made up by quantity. What I'm
saying is that I may not have the "quality" of cracks you have, but I have
a lot more of them. The reason is that I work in the field of pavement
engineering where one of our biggest problem areas is cracking. We can
get, depending on a variety of circumstances, large amounts of fatigue
cracking. As well, especially in Canada and over much of the United States,
we get extensive thermal cracking. The latter cracking problem does not
go away when we resurface or overlay pavements because the underlying
cracks very quickly "reflect" up through the new layer. Our best efforts
to develop stress absorbing membrane interlayers have not yet solved this

problem, and our modelling of the cracking propagation phenomenon involved
is not yet very well developed. Information presented in conferences of
this sort will undoubtedly help us in understanding the fundamental
mechanisms involved.

The three types of cracking problems I have mentioned represent
an extremely large economic loss in terms of accelerated deterioration
and shortened lives of these structures. Solving these problems presents
a major challenge and perhaps I might challenge some of the participants
in this conference to consider them. We need all the help we can get and
certainly the expertise here is elastoplasticity and viscoelasticity
would be valuable. Another valuable aspect represented in this conference
is the concentration on modelling and understanding crack tip mechanics.
There is an unfortunate tendency in my field, and perhaps in others, to
try things to see if they work without first properly understanding or
modelling the mechanisms involved.

Turning to the content of this conference itself, I have not
had the opportunity to read all the papers; indeed many of them are beyond
my expertise. However, what I have read, and the overall impression I
get from reading at least the abstracts of the others, is that a very high
level of scientific and engineering quality is exhibited by the various
contributions. The general lectures are of considerable interest because
they sort of dimension the problem area, and suggest directions for future
work and for problems still to be solved. Then the basic problems session
provides a foundation for the session papers which follow. These session
papers cover a considerable range of particular research topics that the
authors have been involved with. Taken as a whole, the contributions of
the general lectures, the basic problems session and the general session
will provide an invaluable documentation of knowledge for current and
future researchers and for students.

While I think I can at least appreciate the information presented,
and some of the suggestions for future work needed, there is a second
challenge I would like to offer the participants of this conference. It
seems to me that in all the efforts to model and understand crack tip
phenomena, which of course is absolutely necessary if good design solutions
are to be developed, there is one aspect that perhaps has not received a
comparative amount of attention. It involves the development of an accom-
panying stochastic theory of crack initiation and propagation. In addition

to the fundamental understanding required, we also have to deal with probabilities of crack occurrence, with the probabilities of how many will
occur, to what extent they will propagate, etc., for the range of conditions
that prevail. Consequently, let me offer a challenge for people of this
conference to develop a comprehensive stochastic theory in this area.

I would like to close by again expressing my appreciation for
the invitation to attend and offer some comments and by again congratulating
the organizers of this conference for such a fine job and the participants
for providing the scientific and intellectual input necessary to making
any conference a success.

R. Haas
Chairman,
Department of Civil Engineering
University of Waterloo

Preface by the Acting Chairman of the Canadian Committee for Research on the Strength and Fracture of Materials

The Tenth Canadian Facture Conference, held at the University of Waterloo, was the most recent of a series of conferences organized or sponsored by the Canadian Committee for Research on the Strength and Fracture of Materials. It is a pleasure for the Committee to add its congratulations to those of Professor Taplin, to Professor Pindera and Dr. Krasnowski, for the success of this conference.

In his preface, Professor Taplin has indicated how the informal arrangement of national groups interested in strength and fracture has now evolved into a more structured organization, both at the national and international level. Within Canada, our future plans call for one or two local conferences per year (of which CFC10 was an excellent model) as well as continue co-operation and participation in international meetings. In the summer of 1984, two conferences are planned, CFC11 on "Time Dependent Fracture", June 14-15, University of Ottawa; and CFC12 on "Failure Processes in Fibre Composites", May 3-4, Jackson Point, Ontario. Following that, for 1985 a conference on the "Fracture Toughnesses of Weldments" is in the planning stages. The Canadian Committee is also playing an active role in the organization of ICSMA7, the Seventh International Conference on Strength of Metals and Alloys, which is to be held in Montreal, from August 19-23, 1985, under the chairmanship of Professor H. McQueen.

Information on all these conferences as well as other activities of the committee can be obtained from the Secretary of the Organization, Professor M.R. Piggott, c/o Department of Chemical Engineering and Applied Chemistry, University of Toronto, Toronto, Ontario, Canada M5S 1A4.

G.C. Weatherly

Part 1
General Lectures (Invited)

MODELLING PROBLEMS IN CRACK TIP MECHANICS   *Waterloo, Ontario, Canada*
CFC10, University of Waterloo   *August 24-26, 1983*

SOME FINITE ELASTICITY PROBLEMS INVOLVING CRACK-TIPS

Rohan ABEYARATNE
College of Engineering
Michigan State University
East Lansing, Michigan, U.S.A.

1. INTRODUCTION

Crack problems within the theory of linearized elasticity give rise to strain fields which are locally unbounded, even at small values of the applied load. This is, of course, a violation of the approximative assumptions upon which the theory rests. Most early extensions of such studies to beyond the scope of the classical theory of elasticity continue to retain the kinematic assumption of infinitesimal deformations but replace the linear stress-strain law by a nonlinear constitutive relation of some sort. (See for example, the monograph by Hutchinson [1].) The predictions of these analyses cannot possibly be uniformly valid in the immediate vicinity of the crack-tips, no matter how small the loads. It is natural therefore to examine such problems within the setting of a finite theory, which allows for both large deformations and constitutive nonlinearities.

The present paper describes the results of some recent investigations of crack problems within the setting of finite elastostatics. The Mode II and interface-crack problems considered first, serve primarily as illustrative examples to contrast the predictions of the linear and nonlinear theories. The remainder of the paper examines the effect of nonlinearity on the energy released rate - a parameter of some importance in the study of fracture initiation.

The first problem to be considered concerns an infinite slab of a particular homogeneous, isotropic, compressible, nonlinearly elastic material containing a plane crack. The body is loaded in *Mode* II by a remotely applied state of simple shear deformation parallel to the crack-faces. A global solution to the governing plane strain problem is determined and it is observed that the faces of the crack either move apart as a result of this loading and so the crack opens, or there is apparent interpenetration. Moreover in the former case, a tensile stress normal to the crack-faces (which is singular at the crack-tip) is seen to act

at points on the line directly ahead of the crack. This is in contrast
to the corresponding predictions of linear elasticity for this problem,
according to which, see [2], the crack-faces remain contiguous since they
supposedly undergo a purely in-plane (oppositely) directed sliding deformation. Further, the aforementioned normal stress vanishes identically in
the infinitesimal theory. At the end of Section 3.2 we present a physical
argument justifying the presence of these features in the nonlinear problem.

The opening of the Mode II crack was first observed by Stephenson [3]
in his asymptotic study (within the finite theory) of the local fields
near a crack-tip. The issue was subsequently further explored by Knowles
[4] using a somewhat different, but still local, point of view.

We note in passing that a similar phenomenon would, in general, be
expected to occur in the (Mode III) anti-plane shear loading problem
according to the finite theory. As is well-known (e.g., [5]), the imposi-
tion of a purely axial deformation on the boundary of a right cylinder
does not always lead to a state of global anti-plane shear, when nonlinear
effects are taken into account; an interior particle undergoes both out-
of-plane *and* in-plane displacements. Consequently if the body contains a
crack, one expects crack opening, or contact between the crack-faces, to
occur.

The second problem to be considered pertains to an *interface-crack*
between two semi-infinite slabs of dissimilar compressible elastic materials
The corresponding (Mode I, plane strain) problem in linear elasticity has
received repeated attention, primarily in view of the unsatisfactory nature
of its results. As is well-known, e.g. [6,7], the linear theory predicts
the overlapping of the crack-faces, which is of course unacceptable – no
matter how small the interval over which is occurs. A number of recent
investigations have attempted to overcome this difficulty by, for example,
allowing for smooth contact between the crack-faces (e.g., Comninou [8])
or admitting inelastic effects (e.g., Achenbach et al [9]). Recently,
however, Knowles and Sternberg [10] have shown that, as conjectured by
England [7], the unacceptable features here may be attributed solely to
the linearization of the problem; they are found to be absent in the local
solution obtained by these investigators in their asymptotic study of the
plane-stress near-tip deformation field for an interface crack between
two incompressible neo-Hookean sheets. In Section 3.3 of the present paper
we examine the corresponding problem for a crack between two particular

compressible elastic materials within the plane strain setting of finite elasticity. The solution obtained is a global one and, in agreement with [10], predicts the smooth opening of the crack, devoid of interpenetration.

The preceding calculations are carried out within the particular setting of a special compressible elastic material - a member of the class of harmonic materials introduced by John [11]. Despite the lack of an experimental basis, their mathematical amenability makes them an attractive choice for analytical studies in finite elasticity, e.g., [12-14]. They are suitable for our present purposes, since, the aim of the first part here is *merely* to illustrate certain possibilities within the nonlinear theory.

In the second part of this paper we review some recent results pertaining to the *energy release rate* associated with nonlinear crack problems; a more detailed discussion may be found in [15]. If a crack in a deformed solid is to propagate, it requires energy and the so-called "energy release rate" represents the rate at which energy is made available to the crack for this purpose. Experiments carried out by Begley and Landes [16] have demonstrated the potential that this parameter has for use as a fracture initiation criterion. This in turn has prompted many investigations aimed at calculating its value under nonlinear conditions. These have primarily consisted of numerical investigations, as well as analytical approximations based on interpolation. The only exact solution appears to be that due to Amazigo [17] for a pure power-law material under conditions of anti-plane shear.

The well-known small-scale yielding approximation for the energy release rate uses the argument that when the applied loads are small, the nonlinearities in the problem are confined to a small region near the crack-tips and that the linear elastic solution provides a good approximation elsewhere. The approximation used here is a natural extension of this idea and essentially, replaces the linear elastic solution in the preceding argument by an improved approximation which takes into account the nonlinearities of the problem. Consider for example the plane strain Mode I crack problem. In the absence of the crack, the body is in a homogeneous state of (finite) plane-strain uniaxial tension. If one were to assume that the presence of the crack causes only a small perturbation of this homogeneous state, one can then carry out an analysis based on the theory of small deformations superposed on a large deformation. The

results based on such an approximation would clearly be invalid in the
vicinity of the crack, but one would expect it to provide a reasonable
approximation at points distant from the crack-tips. Since our interest
lies *solely* in estimating the energy release rate, and since this can be
written in the form of a line integral J taken over a contour that is *far*
from the crack tips over most of its length, (Rice [2]), one expects such
an approximation to be suitable for our purposes, at least when confined
to moderate levels of loading.

In this paper we summarize such an analysis for determining the energy
release rate associated with the Mode I and Mode III crack problems. The
plane strain calculations are carried out for a general elastic material
which is homogeneous, isotropic and incompressible; the Mode III calculations are appropriate for any material in this same class provided, of
course, it admits a nontrivial state of anti-plane shear. The results
obtained in the anti-plane shear case are compared with Amazigo's exact
solution [17] in the particular case of a pure power-law material. The
plane strain solution is similarly compared with the finite element results
given by He and Hutchinson [18]. In both cases, the agreement is found
to be surprisingly good.

2. PRELIMINARIES FROM FINITE ELASTICITY

Consider an elastic body which occupies a three-dimensional region
R_0 in its unstressed state. The body is subjected to a deformation which
carries the particle located at x in the undeformed body to the new
location $y = \hat{y}(x) = x+u(x)$. Let R denote the region occupied by the body
in the deformed equilibrium configuration. The deformation gradient[1]
tensor F, left Cauchy-Green deformation tensor G and Eulerian strain tensor E are given by

$$F = \nabla\hat{y} , \qquad G = FF^T , \qquad E = (1-G^{-1})/2 . \tag{2.1}$$

The deformation from R_0 to R is one to one and the associated Jacobian
determinant J is assumed to be positive,

$$J = \det F > 0 . \tag{2.2}$$

The fundamental scalar invariants I_1, I_2, I_3 of G are

$$I_1 = \operatorname{tr} G , \qquad I_2 = [(\operatorname{tr}G)^2-(\operatorname{tr}G^2)]/2 , \qquad I_3 = \det G = J^2 , \tag{2.3}$$

---

[1] An extensive discussion of finite elasticity may be found in [19].

while the principal stretches λ_1, λ_2, λ_3 (> 0) of the deformation are the
eigenvalues of $\sqrt{G}$. In the particular case of an incompressible material,
only locally volume preserving deformations are admissible, whence
$I_3 = J = 1$.

Next, let $\sigma(x)$ and $\tau(y)$ be the first Piola-Kirchhoff (or nominal) and
Cauchy (or true) stress tensor fields. The components of σ and τ represent forces per unit undeformed and deformed areas respectively and they
are related by

$$\sigma = J \tau F^{-T} . \tag{2.4}$$

The conditions for equilibrium, in the absence of body force, are

$$\text{div } \tau(y) = 0 , \quad \tau = \tau^T \text{ on R} , \tag{2.5}$$

or equivalently

$$\text{div } \sigma(x) = 0 , \quad \sigma F^T = F\sigma^T \text{ on } R_0 . \tag{2.6}$$

The true traction t acting on a surface S in the deformed body is given
by $t = \tau n$ while the nominal traction s obeys $s = \sigma N$. The vectors n and
N here are the unit normals to S and S_0 respectively, where S_0 is the
surface in the undeformed configuration which is carried into S by the
deformation. In the particular case when S is free of traction, one has
$t = 0$ on S and $s = 0$ on S_0.

Suppose that the constitutive behaviour of the homogeneous elastic
body at hand is characterized by its strain energy density per unit undeformed volume $W(F)$. If the material is isotropic, W depends on F only
through the invariants I_1, I_2, I_3: $W = W(I_1, I_2, I_3)$ for a compressible
material and $W = W(I_1, I_2)$ for an incompressible one. The stress is then
related to the deformation, for a compressible material, through

$$\sigma = 2 \left(\frac{\partial W}{\partial I_1} + I_1 \frac{\partial W}{\partial I_2} \right) F - 2 \frac{\partial W}{\partial I_2} GF + 2I_3 \frac{\partial W}{\partial I_3} F^{-T} , \tag{2.7}$$

and for an incompressible one through

$$\sigma = 2 \left(\frac{\partial W}{\partial I_1} + I_1 \frac{\partial W}{\partial I_2} \right) F - 2 \frac{\partial W}{\partial I_2} GF - pF^{-T} . \tag{2.8}$$

Corresponding expressions for the true stress τ may be obtained from (2.4),
(2.7), (2.8).

We now consider certain special states of deformation. In a *simple shear deformation* with amount of shear k, one has $u_1 = u_2 = 0$ and $u_3 = kx_3$. Thus, according to (2.1)-(2.4), (2.7), (2.8), the true shear stress component τ_{12} associated with this deformation is given by

$$\tau_{12} = \tau(k) \equiv 2k \left(\frac{\partial W}{\partial I_1} + \frac{\partial W}{\partial I_2} \right) , \qquad I_1 = I_2 = 3+k^2 , \qquad I_3 = 1 . \qquad (2.9)$$

The corresponding normal stress components τ_{11}, τ_{22}, τ_{33} (which do not, in general, vanish in the finite theory - the so-called "Poynting effect") - may also be calculated from the aforementioned equations.

Next, suppose that the region R_o is cylindrical with its generators parallel to the x_3-axis, and that the body is subjected to a plane deformation. Let D_o denote the cross-section of R_o in the plane $x_3 = 0$. The components of displacement associated with a state of *plane strain* are $u_1 = u_1(x_1,x_2)$, $u_2 = u_2(x_1,x_2)$, $u_3 = 0$. For an isotropic, *compressible* material under these conditions, one has

$$I_1 = I+1 , \qquad I_2 = I+J^2 , \qquad I_3 = J^2 \text{ where } I = \text{tr}(\underset{\sim}{G})-1 , \qquad (2.10)$$

so that $W = W(I_1,I_2,I_3) = W(I,J)$. It can be shown that the in-plane stress deformation relations in this case simplify to

$$\underset{\sim}{\sigma} = 2 \frac{\partial W}{\partial I} \underset{\sim}{F} + J \frac{\partial W}{\partial J} \underset{\sim}{F}^{-T} , \qquad \underset{\sim}{\tau} = \frac{2}{J} \frac{\partial W}{\partial I} \underset{\sim}{G} + \frac{\partial W}{\partial J} \underset{\sim}{1} . \qquad (2.11)$$

Moreover, the in-plane equilibrium equations (2.6) hold if and only if the components of $\underset{\sim}{\sigma}$ are expressed in terms of two stress-functions $\psi_1(x_1,x_2)$, $\psi_2(x_1,x_2)$ by

$$\sigma_{11} = \partial\psi_2/\partial x_2 , \qquad \sigma_{12} = -\partial\psi_2/\partial x_1,$$
$$\sigma_{21} = -\partial\psi_1/\partial x_2 , \qquad \sigma_{22} = \partial\psi_1/\partial x_1. \qquad (2.12)$$

On the other hand, for an isotropic *incompressible* material under *plane strain* conditions, one finds that

$$I_1 = I_2 = I+1 , \qquad I = \text{tr}(\underset{\sim}{G})-1 , \qquad (2.13)$$

and so $W = W(I_1,I_2) = W(I)$. The in-plane constitutive law here is

$$\underset{\sim}{\sigma} = 2W'(I)\underset{\sim}{F} - p\underset{\sim}{F}^{-T} , \qquad \underset{\sim}{\tau} = 2W'(I)\underset{\sim}{G}-p\underset{\sim}{1} . \qquad (2.14)$$

It is useful to note that any plane deformation of an incompressible material can be viewed locally as a simple shear deformation in a suitable direction with an amount of shear k_e,

$$k_e = \sqrt{I-2} , \qquad I = tr(\underset{\sim}{G})-1 , \tag{2.15}$$

followed (or preceded) by a suitable rotation, [20].

Finally, consider the displacement field $u_1 = u_2 = 0$, $u_3 = u(x_1,x_2)$ corresponding to a state of *anti-plane strain* of the cylindrical body. Not all elastic materials can sustain such a deformation and in what follows, it will be tacitly assumed that we are only considering those incompressible materials which admit nontrivial states of anti-plane shear, Knowles [5,21]. Here, one has $I_1 = I_2 = I$ where $I = 3+|\nabla u|^2$, so that $W = W(I_1,I_2) = W(I)$. The governing field equations now reduce to the single differential equation

$$div(W'(3+|\nabla u|^2) \ grad \ u) = 0 \ on \ D_o . \tag{2.16}$$

The corresponding stress components are $(\alpha,\beta = 1,2)$

$$\tau_{3\alpha} = \tau_{\alpha 3} = \sigma_{3\alpha} = \sigma_{\alpha 3} = 2W'(3+|\nabla u|^2) \partial u/\partial x_\alpha , \tag{2.17}$$

$$\tau_{\alpha\beta} = \sigma_{\alpha\beta} = \sigma_{33} = 0 , \qquad \tau_{33} = 2W'(3+|\nabla u|^2)|\nabla u|^2 . \tag{2.18}$$

3. THE EFFECT OF NONLINEARITY ON THE DEFORMATION OF A CRACK: TWO EXAMPLES

In this section we summarize the results of two plane strain crack problems for an elastic body composed of a particular (compressible) harmonic material.

3.1 Harmonic Materials. Solution of the Field Equations

The class of harmonic materials introduced by John [11] are characterized by the plane strain elastic potential

$$W = 2\mu\{H(R)-J\} , \qquad R = (I+2J)^{1/2} = \lambda_1+\lambda_2 > 0 , \tag{3.1}$$

where $\mu(> 0)$ denotes the infinitesimal shear modulus and H is a constitutive function defined for all positive values of its argument.

The equations governing plane deformations of a harmonic material take on a particularly simple form when cast in terms of complex-valued functions, e.g., [12-14]. To this end let $w(x_1,x_2)$ and $\psi(x_1,x_2)$ be the complex-valued functions defined for all $(x_1,x_2) \in D_o$ by

$$w(x_1,x_2) = \hat{y}_1(x_1,x_2) + i\ \hat{y}_2(x_1,x_2)\ ,$$

$$\psi(x_1,x_2) = \psi_1(x_1,x_2) + i\ \psi_2(x_1,x_2)\ . \tag{3.2}$$

Here $\hat{y}_1$, $\hat{y}_2$ are the components of the position vector after deformation, and ψ_1, ψ_2 are the stress functions introduced in (2.12). Then, as shown for example by Ogden and Isherwood [13], the governing field equations are equivalent to

$$\psi + 2\mu\ w = 2\mu\ A(z) \quad \text{on } D_0\ , \tag{3.3}$$

$$\partial\psi/\partial x_1 - i\ \partial\psi/\partial x_2 = 2\mu(2H'(R)-R)\ \frac{A'(z)}{|A'(z)|} \quad \text{on } D_0\ , \tag{3.4}$$

with $H'(R) = |A'(z)|$. Here $A(z)$ is an arbitrary analytic function of $z = x_1+ix_2$ on D_0 and $A'(z)$ has no zeros on D_0. Alternatively, Varley and Cumberbatch [14] demonstrate that if the (nominal) stress-stretch response of the harmonic material in (plane) uniaxial tension is denoted by $\sigma = \hat{\sigma}(\lambda)$ for $\lambda > 0$, then equation (3.4) can be written in the form

$$\partial\psi/\partial x_1 - i\ \partial\psi/\partial x_2 = \hat{\sigma}(|A'(z)|)\ \frac{A'(z)}{|A'(z)|}\ . \tag{3.5}$$

The solution of the governing field equations thus reduces to the integration of (3.5) for the stress-function ψ. Equation (3.3) then yields the "deformation" w.

In order to facilitate the closed-form integration of (3.5), we choose a particular (convenient) form for the response function $\hat{\sigma}(\lambda)$ in uniaxial tension:

$$\sigma = \hat{\sigma}(\lambda) \equiv 2\mu(\lambda-\lambda^{-1}) \quad \text{for} \quad \lambda > 0\ , \tag{3.6}$$

which is a particular member of the special class of harmonic materials considered by Varley and Cumberbatch (see equation (7.1) of [14]). Attention here is restricted to this one member, since the others in [14] behave unrealistically (in the transverse direction) at sufficiently severe levels of plane uniaxial tension (which is precisely the state of interest at points on a traction-free crack-face close to a crack-tip). On confining attention to the particular constitutive law (3.6), equation (3.5) may be integrated to give

$$\psi(x_1,x_2) = \mu\{A(z) - \overline{C(z)} + (\bar{z}-z)/\overline{A'(z)}\}\ , \tag{3.7}[2]$$

which together with (3.3) yields

$$w(x_1,x_2) = \frac{1}{2}\{A(z) + \overline{C(z)} + (z-\bar{z})/\overline{A'(z)}\}\ . \tag{3.8}$$

[2]A bar denotes the complex congugate.

The functions $A(z)$ and $C(z)$, analytic on D_0, are to be determined from the boundary conditions.

3.2 Mode II Crack Problem

The cross-section D_0 of the undeformed body containing a plane crack of length $2a$, as well as the coordinate system, is shown in Figure 1. The body is subjected to a remotely applied in-plane simple shear deformation $y_1 \sim x_1 + k_\infty x_2$, $y_2 \sim x_2$ as $x_1^2 + x_2^2 \to \infty$, so that according to $(3.2)_1$

$$w \sim (1 - i \tfrac{k_\infty}{2}) z + (i \tfrac{k_\infty}{2}) \, \bar{z} \quad as \quad |z| \to \infty . \tag{3.9}$$

The amount of shear applied at infinity has been denoted by k_∞. The deformed faces of the crack are presumed to be free of traction so that in view of (2.12), (3.2) and the discussion following (2.6), the stress-function ψ must vanish on the crack-faces:

$$\psi(x_1, \; 0\pm) = 0 \quad for \quad -a < x_1 < a . \tag{3.10}$$

Finally, the components of displacement, and hence $w(x_1, x_2)$, are required to be bounded at the crack-tips.

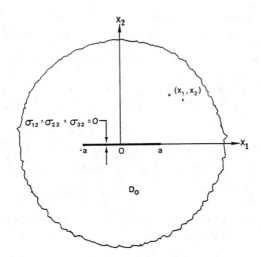

Figure 1 - Cross-Section of Undeformed Body Containing a Crack

On using (3.7) and (3.8), the preceding requirements on w and ψ may be recast in terms of the analytic functions $A(z)$ and $C(z)$. Equation (3.9) then leads to

$$A(z) \sim \alpha z , \qquad C(z) \sim \gamma z \qquad \text{as } |z| \to \infty , \tag{3.11}$$

where the constants α and γ obey

$$\alpha + 1/\overline{\alpha} = 2 - ik_\infty , \qquad \overline{\gamma} - 1/\overline{\alpha} = ik_\infty . \tag{3.12}$$

Similarly, (3.10) yields

$$A(z) = \overline{C(z)} \qquad \text{for} \qquad x_2 = 0\pm , \quad |x_1| < a . \tag{3.13}$$

The Mode II crack problem thus reduces to the determination of two functions, $A(z)$ and $C(z)$, analytic on D_0, bounded at the crack-tips and conforming to (3.11) - (3.13). This is a standard problem in analytic function theory and may be solved by first transforming (3.13) into a Riemann-Hilbert problem. The details of this procedure may be found in England [22]. The resulting solution is

$$A(z) = z + \beta\sqrt{z^2 - a^2} , \tag{3.14}$$

$$C(z) = z - \overline{\beta}\sqrt{z^2 - a^2} , \tag{3.15}$$

with the constant β given by

$$\beta = \delta - \frac{ik_\infty}{2}(\delta + 1) , \qquad \delta = \pm \frac{|k_\infty|}{\sqrt{4 + k_\infty^2}} . \tag{3.16}[3]$$

The square root $\sqrt{(z^2 - a^2)}$ here, and henceforth, refers to the principal branch.[4]

Details of the corresponding deformation and stress fields may now be examined. For example, it follows from (3.14), (3.15), (3.8) and (3.2) that the *deformed shape* of the upper crack-face is given by

$$y_1 = x_1 + \frac{k_\infty}{2}(\delta + 1)\sqrt{a^2 - x_1^2} , \qquad y_2 = \delta\sqrt{a^2 - x_1^2} , \tag{3.17}$$

for $|x_1| < a$, so that the crack opens if $\delta > 0$ into an elliptical shape.[5] Interpenetration is predicted when $\delta < 0$ in which case the solution here is of course invalid. In *neither of these cases* is this a purely sliding

---

[3] The ambiguity in the sign of δ is discussed in Section 3.4.

[4] $\sqrt{(z^2 - a^2)} = \sqrt{r_1 r_2} \exp i(\theta_1 + \theta_2)/2$ where $x_1 = a + r_1\cos\theta_1 = r_2\cos\theta_2 - a$, $x_2 = r_1\sin\theta_1 = r_2\sin\theta_2$, $r_1 > 0$, $r_2 > 0$, $-\pi < \theta_1, \theta_2 \leq \pi$.

[5] When $\delta > 0$ one can verify from (3.14), (3.16) that, as required, $A'(z)$ has no zeros in the open region D_0.

deformation of the crack-faces as predicted by the linear theory. Somewhat surprisingly, the line $x_2 = 0$, $x_1 > a$ ahead of the crack is seen to remain undistorted ($y_1 = x_1$, $y_2 = 0$) by the deformation.

The stress components σ_{12}, σ_{22} ahead of the crack (on $x_2 = 0$, $x_1 > a$) are found from (2.12), (3.2), (3.7), (3.14) and (3.15) to be

$$\sigma_{12} = \mu k_\infty (\delta+1) \ x_1/\sqrt{x_1^2 - a^2} \ , \qquad \sigma_{22} = 2\mu\delta \ x_1/\sqrt{x_1^2 - a^2} \ . \qquad (3.18)$$

The *presence of a normal stress* σ_{22} along this line is again contrary to the corresponding result according to the linear theory. It should be noted that the parameter δ, and so the nonlinear effects here, are of the first order in the amount of applied shear k_∞ (as $k_\infty \to 0$). This is peculiar to the special constitutive relation (3.6) and in general, one would expect this dependency to be of the second - or higher - order.

Finally, we note that it is not difficult to provide a heuristic explanation for the aforementioned features based on the Poynting effect. Recall that a state of simple shear, according to the finite theory, requires the presence of both normal and shear stresses. Indeed, in the Mode II crack-problem here, one finds from (2.12), (3.2), (3.7), (3.14) and (3.15) that the normal stress σ_{22} at points far from the crack obeys

$$\sigma_{22} \to 2\mu \ \delta \qquad \text{as} \qquad x_1^2 + x_2^2 \to \infty \ . \qquad (3.19)$$

Thus, when $\delta > 0$, there is a tensile stress acting in a direction normal to the crack-faces at infinity, which naturally leads to crack opening and a (singular) tensile normal stress σ_{22} ahead of the crack. On the other hand when $\delta < 0$, the normal stress at infinity is compressive which ought to result in contact between the crack faces.[6]

3.3 Interface - Crack Problem

Suppose now that the body under consideration consists of two semi-infinite blocks of different compressible elastic materials which are bonded along their common interface, the (x_1,x_3)-plane, except for a through crack of length 2a. Figure 2 shows the cross-section of the undeformed body in the plane $x_3 = 0$ with the upper and lower open half-planes being denoted by D_1 and D_2. The blocks are presumed to be composed of the

---

[6]It should be noted however that in an earlier investigation [4], Knowles observed that, for a more general class of materials, there is no direct correlation between the Poynting effect and the opening (or closing) of the crack, at least in the asymptotic sense of his investigation.

particular harmonic material introduced previously by (3.6) but with different shear moduli μ_1, μ_2 respectively. The formulation presented in
Section 3.1 continues to hold in the present case and one finds that in
any plane deformation, the complex-valued stress function ψ and deformation w are given by

$$w(x_1,x_2) = \frac{1}{2} \{A_\alpha(z) + \overline{C_\alpha(z)} + (z-\bar{z})/\overline{A'_\alpha(z)}\} \qquad \text{on} \quad D_\alpha \,, \qquad (3.20)$$

$$\psi(x_1,x_2) = \mu_\alpha \{A_\alpha(z) - \overline{C_\alpha(z)} + (\bar{z}-z)/\overline{A'_\alpha(z)}\} \qquad \text{on} \quad D_\alpha \,, \qquad (3.21)$$

($\alpha = 1,2$). The four functions $A_\alpha(z)$, $C_\alpha(z)$ are analytic on D_α, $A'_\alpha(z)$ has
no zeros in D_α and $A_\alpha(z)$, $C_\alpha(z)$ are bounded at the crack-tips. They are
to be determined through the boundary conditions.

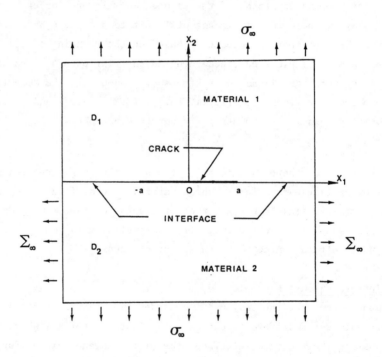

Figure 2 - Interface Crack Between Two Semi-Infinite Elastic Slabs

Suppose that this body is subjected to a remotely apply tensile stress
σ_∞ (> 0) normal to the crack-faces. As pointed out by Rice and Sih [23]
such a loading, in the absence of any other applied stresses, is inconsis-
tent with the bond conditions. The loading may be made consistent with

the presumed state of plane strain, by applying, for example, a suitable
additional normal stress $\sigma_{11} = \Sigma_\infty$ on the lower half plane. In the present problem the value of Σ_∞ required for this purpose is

$$\Sigma_\infty = \frac{\mu_1 - \mu_2}{\mu_1} \sigma_\infty \left(1 + \frac{\sigma_\infty}{2\mu_2 \lambda_\infty}\right)^{-1} , \qquad \sigma_\infty = 2\mu_1 (\lambda_\infty - \lambda_\infty^{-1}) . \tag{3.22}$$

On using (2.12) and (3.2) we may write these prescribed load conditions as

$$\psi \sim \frac{\sigma_\infty}{2} (z + \bar{z}) \quad \text{as} \quad |z| \to \infty, \quad x_2 > 0 , \tag{3.23}$$

$$\psi \sim \left(\frac{\sigma_\infty + \Sigma_\infty}{2}\right) z + \left(\frac{\sigma_\infty - \Sigma_\infty}{2}\right) \bar{z} \quad \text{as} \quad |z| \to \infty , \quad x_2 < 0 . \tag{3.24}$$

If the faces of the deformed crack are free of traction, then one re-
quires that ψ vanish on $x_2 = 0\pm$, $|x_1| < a$. In view of (3.21) this implies
that

$$A_1(z) = \overline{C_1(z)} \quad \text{on} \quad x_2 = 0+ , \quad |x_1| < a ,$$
$$A_2(z) = \overline{C_2(z)} \quad \text{on} \quad x_2 = 0- , \quad |x_1| < a . \tag{3.25}$$

Turning to the bond conditions, one must have the continuity of displacements, and so of w, across the bonded interface. In view of (3.20),
this requires that for $|x_1| > a$,

$$\lim_{x_2 \to 0+} \{A_1(z) + \overline{C_1(z)}\} = \lim_{x_2 \to 0-} \{A_2(z) + \overline{C_2(z)}\} . \tag{3.26}$$

Similarly, the continuity of traction across the bond implies that ψ is
continuous across the interface and so from (3.21) one has

$$\lim_{x_2 \to 0+} \mu_1 \{A_1(z) - \overline{C_1(z)}\} = \lim_{x_2 \to 0-} \mu_2 \{A_2(z) - \overline{C_2(z)}\} , \quad |x_1| > a . \tag{3.27}$$

The solution of the interface-crack problem is thus reduced to deter-
mining four analytic functions $A_\alpha(z)$, $C_\alpha(z)$ on D_α which conform to the
requirements (3.22) - (3.27) and are bounded at the crack-tips. We again
omit the details of the procedure for determining these functions. The
main step involves the introduction of two auxiliary functions $\Omega(z)$ and
$\theta(z)$ which are analytic on the entire z-plane cut along the crack, and
are defined by

$$\Omega(z) = A_\alpha(z) - \overline{C_\beta(\bar{z})} \quad \text{on} \quad D_\alpha ,$$
$$\theta(z) = \mu_\alpha A_\alpha(z) + \mu_\beta \overline{C_\beta(\bar{z})} \quad \text{on} \quad D_\alpha , \tag{3.28}$$

with (α, β) taking the values $(1,2)$ and $(2,1)$ in each of (3.28). This leads to a standard problem for $\Omega(z)$ and $\theta(z)$ (see England [22]). Thus, one finds that

$$(\mu_1 + \mu_2) \, A_\alpha(z) = \theta(z) + \mu_\beta \, \Omega(z) \, , \tag{3.29}$$

$$(\mu_1 + \mu_2) \, C_\alpha(z) = \overline{\theta(\bar{z})} - \mu_\beta \, \overline{\Omega(\bar{z})} \, , \tag{3.30}$$

where $(\alpha, \beta) = (1,2)$ and $(2,1)$, and

$$\theta(z) = \alpha z \, , \qquad \Omega(z) = \gamma \sqrt{z^2 - a^2} \, . \tag{3.31}$$

The constants α and γ (both > 0) are given by

$$\sigma_\infty = 2\mu_1 \left(\frac{\alpha}{\mu_1 + \mu_2} - \frac{\mu_1 + \mu_2}{\alpha} \right) , \tag{3.32}$$

$$\gamma = \frac{\sigma_\infty}{2} \left(\frac{\mu_1 + \mu_2}{\mu_1 \mu_2} \right) . \tag{3.33}$$

It may be verified that $A'_\alpha(z)$ has no zeros on D_α as required.

The corresponding deformation and stress fields may now be readily calculated from (3.29) - (3.33), (3.20), (3.21), (3.2) and (2.12). Here, we are primarily interested in the *deformed shape* of the crack. This is found to be described by

$$y_1 = \frac{\alpha}{\mu_1 + \mu_2} \, x_1 \, , \qquad y_2 = \frac{\mu_2 \gamma}{\mu_1 + \mu_2} \, \sqrt{a^2 - x^2} \, , \qquad |x_1| < a \, , \tag{3.34}$$

for the upper face of the crack and by

$$y_1 = \frac{\alpha}{\mu_1 + \mu_2} \, x_1 \, , \qquad y_2 = - \frac{\mu_1 \gamma}{\mu_1 + \mu_2} \, \sqrt{a^2 - x^2} \, , \qquad |x_1 \sqrt{} < a \, , \tag{3.35}$$

for the lower face. Since α and γ are both positive it is clear that according to the finite theory, the interface crack here opens, and *does not involve interpenetration*. This is in contrast to the corresponding plane strain interface crack problem in linear (compressible) elasticity. It is apparent from (3.34), (3.35) that, in fact, the interface crack here opens smoothly, Figure 3.

The interface $x_2 = 0$, $x_1 > 0$, ahead of the crack continues to remain straight after deformation:

$$y_1 = \frac{\alpha}{\mu_1 + \mu_2} \, x_1 \, , \qquad y_2 = 0 \, , \qquad x_1 > a \, . \tag{3.36}$$

The nominal stress components σ_{22}, σ_{12} along the bond are found to be

$$\sigma_{22} = \frac{\sigma_\infty x_1}{\sqrt{x_1^2 - a^2}} \quad , \qquad \sigma_{12} = 0 \; . \tag{3.37}$$

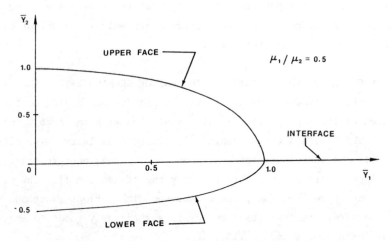

Figure 3 - Deformed Crack-Faces and Interface

3.4 Concluding Remarks

Here we wish to draw attention to two shortcomings in the solutions summarized previously, both of which may be traced back to the constitutive behaviour of the specific harmonic material utilized in the analysis.

The specific material which has been considered here is compressible, as may be verified by examining its response in a state of plane isotropic extension. However its Poisson's ratio ν has the value 1/2 (which merely implies that the volume changes at infinitesimal deformations are of the second - or higher - order in the displacement gradients). Consequently one might argue that the results for the compressible material here ought to be compared with those for an incompressible material in the linear theory, in which case, neither theory predicts interpenetration in the interface-crack problem.

Next, while it was convenient for our purposes to characterize a harmonic material through its uniaxial stress-stretch law, the strain energy function W (or the constitutive function H) associated with the specific response (3.6) used by us (and in [14]) turns out to be multi-valued in general. [One can avoid this issue by using an alternative formulation based on a (single-valued) complementary energy function. Also, this multi-valuedness is essentially the reason for the ambiguity in the sign of δ in Section 3.2].

We note however that these shortcomings may both be overcome by picking the branch[7] of W pertaining to large deformations and viewing the solutions here as being purely local, appropriate only in an asymptotic sense in the vicinity of the crack-tips.

4. THE EFFECT OF NONLINEARITY ON THE ENERGY RELEASE RATE

In this section we summarize certain results pertaining to the calculation of the energy release rate associated with the Mode I (plane strain) and Mode III crack-problems in finite elasticity. The class of materials considered is isotropic, homogeneous and incompressible but is otherwise essentially unrestricted (see end of Section 1). The calculation is an approximate one, but comparison with various exact and numerical solutions indicate that its accuracy is surprisingly good. A more detailed discussion may be found in [15].

4.1 Anti-Plane Shear: Mode III

The cross-section D_0 of the infinite elastic body here is shown in Figure 1. The body is subjected to a remotely applied simple shear deformation in the x_3-direction (parallel to the crack-faces) while the crack remains free of traction. The resulting deformation is assumed to be one of anti-plane shear (see discussion following (2.15) in Section 2).

The mathematical problem consists of determining the out-of-plane displacement $u(x_1,x_2)$ satisfying the differential equation (2.16) as well as the boundary conditions

$$\tau_{32} = 2W'(3+|\nabla u|^2)\, u_{,2} = 0 \quad \text{on} \quad x_2 = 0\pm\,, \quad |x_1| < a\,,$$

$$u \sim k_\infty x_2 \quad \text{as} \quad x_1^2 + x_1^2 \to \infty\,. \tag{4.1}[8]$$

Here k_∞ is the amount of shear applied at infinity.

---

[7] i.e., $2H'(R) = R+\sqrt{(R^2-4)}$ for "large" values of R. Observe that the value of the Poisson's ratio ν is arbitrary here, since it is associated with the constitutive behaviour at "small" values of R ($R\to2$).

[8] A comma followed by a subscript denotes differentiation with respect to the corresponding x-coordinate.

The energy release rate J associated with this problem is

$$J = \int_C \{W(I)N_1 - 2W'(I)u_{,\alpha} u_{,1} N_\alpha\}\, ds, \qquad I = 3+|\nabla u|^2 , \qquad (4.2)$$

(Rice [2]), where C is any simple closed curve which encloses the right crack-tip in its interior but does not include the left one and N_α are the components of the unit outward normal to C. Our aim here is to estimate the value of J.

In the absence of the crack, the deformation of the body is one of simple shear and the associated displacement field is $u = k_\infty x_2$. If we assume that the presence of the crack causes only a small disturbance to this crack-free equilibrium state, we would have

$$u(x_1,x_2) = k_\infty x_2 + \tilde{u}(x_1,x_2) , \qquad |\nabla \tilde{u}| \ll 1 \quad \text{on} \quad D_0 . \qquad (4.3)$$

While such a hypothesis is undoubtedly invalid in the vicinity of the crack, it seems reasonable to assume that it would be true at points distant from the crack-tips. Since our interest here lies solely in the computation of the path-independent integral J, and since for this purpose we may consider a path which is essentially far removed from the crack-tips, the errors introduced by such an assumption are probably small, at least when confined to moderate levels of loading.

Accordingly we now assume that the displacement field u is of the form given by (4.3). Linearization of (2.17) based on a small displacement gradient $\nabla \tilde{u}$ leads to

$$\tau_{31} = \mu(k_\infty)\tilde{u}_{,1} , \qquad \tau_{32} = \tau_\infty + \tau'(k_\infty)\tilde{u}_{,2} , \qquad (4.4)$$

where τ_∞ is the shear stress τ_{32} at infinity, $\tau_\infty = \tau(k_\infty)$. Moreover, we have denoted the secant, tangent and infinitesimal shear moduli of the material (all > 0) by

$$\mu(k) = \frac{\tau(k)}{k} , \qquad \tau'(k) = \frac{d\tau(k)}{dk} , \qquad \mu_0 = \tau'(0) , \qquad (4.5)$$

respectively, $\tau(k) = 2kW'(3+k^2)$. The nonlinear boundary-value problem (2.16), (4.1) may now be replaced by the linear problem

$$\mu(k_\infty)\tilde{u}_{,11} + \tau'(k_\infty)\tilde{u}_{,22} = 0 \quad \text{on} \quad D_0 , \quad \tilde{u}_{,2} = -\frac{\tau_\infty}{\tau'(k_\infty)} \quad \text{on } x_2 = 0\pm, \ |x_1| < a ,$$

$$\qquad (4.6)$$

$$\tilde{u} \sim 0 \qquad \text{as} \qquad x_1^2 + x_2^2 \to \infty .$$

Similarly, the linearization of the integrand of the J-integral (4.2) yields

$$J = \int_C \frac{1}{2} \{-\mu(k_\infty)\tilde{u}^2_{,1}N_1 + \tau'(k_\infty)\tilde{u}^2_{,2}N_1 - 2\tau'(k_\infty)\tilde{u}_{,1}\tilde{u}_{,2}N_2\}ds , \qquad (4.7)$$

to leading order. Solution of the linear boundary-value problem (4.6) gives $\tilde{u}$ when then permits the computation of J according to (4.7).

This end is most conveniently achieved by a rescaling which transforms the first of (4.6) into Laplace's equation. The resulting problem is now similar to that according to the linearized theory of elasticity whence

$$\tilde{u}(x_1,x_2) = (\tau_\infty/\sqrt{\mu\tau'}) \; Im \; \{\sqrt{z^2-a^2} - z\} , \qquad (4.8)$$

where $z = x_1 + sx_2$, $s = i\sqrt{(\mu/\tau')}$. The value of J may then be readily calculated from (4.7) to be given by *the simple formula*

$$J = \frac{\pi a}{2} \frac{\tau^2_\infty}{\{\tau'(k_\infty)\mu(k_\infty)\}^{1/2}} . \qquad (4.9)$$

In the particular case when the applied load is small, the tangent and secant moduli coincide with the infinitesimal shear modulus μ_0 and (4.9) reduces to

$$J = \frac{\pi a}{2\mu_0} \tau^2_\infty , \qquad (4.10)$$

which is, of course, identical to the result for J according to the small scale yielding approximation of fracture mechanics.

Next consider the pure power-law material studied by Amazigo [17],

$$\tau(k) = \tau_0 \left(\frac{k}{\alpha k_0} \right)^{1/n} . \qquad (4.11)$$

In this case (4.9) yields

$$J = \frac{\pi a}{2} \tau_\infty k_\infty \sqrt{n} . \qquad (4.12)$$

Figure 4 shows the variation of J with the material hardening exponent n according to (4.12). The exact values shown there are taken from Table 1 of [17]. The comparison is seen to be remarkably good over a wide range of values of the material exponent n, with the formula (4.12) becoming less accurate as the shear response curve becomes "flatter", $n \to \infty$. It

should be noted that the comparison made here is in fact for *all* values
of the applied load and not merely for moderate ones.

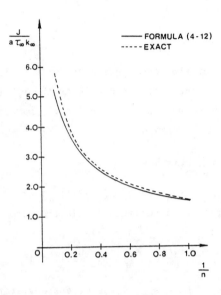

*Figure 4 - Variation of the Energy Release Rate J with the Material
Hardening Exponent n in Anti-Plane Shear. Comparison of
Formula (4.13) with Exact Results [17].*

4.2 Plane Strain: Mode I

The energy release rate associated with the Mode I crack-problem can
be estimated through a procedure entirely analogous to that just described.
The calculations are somewhat more tedious than in the Mode III case and
are omitted here.   In the present case one finds that

$$J = \frac{\pi a}{2} \frac{\sigma_\infty^2}{\{\tau'(k_\infty)\mu(k_\infty)\}^{1/2}} \cdot \frac{1}{(1+\lambda_\infty^2)} \cdot \left\{1 - \left(\frac{1-\lambda_\infty^{-2}}{\lambda_\infty+\lambda_\infty^{-1}}\right)^2 \frac{\mu(k_\infty)}{\tau'(k_\infty)}\right\}^{-1}, \qquad (4.13)$$

where λ_∞ (> 1) is the principal stretch corresponding to the remotely
applied true stress σ_∞, $\sigma_\infty = \mu(k_\infty)(\lambda_\infty^2-\lambda_\infty^{-2})$, and $k_\infty \equiv \lambda_\infty-\lambda_\infty^{-1}$. Recall
from Section 2 that any plane strain deformation of an incompressible
material can always be decomposed locally into a rigid rotation followed,
or preceded, by a simple shear.   The in-plane behaviour of the material is
therefore essentially determined by its response in simple shear, $\tau = \tau(k)$;
$\mu(k)$ and $\tau'(k)$ in (4.13) denote the secant and tangent moduli (4.5) of
the material in shear.

In the particular case when the applied load is small, the tangent and secant moduli coincide with the infinitesimal shear modulus μ_0 and (4.13) reduces to the classical result

$$J = \frac{\pi a}{4\mu_0} \sigma_\infty^2 . \tag{4.14}$$

Suppose instead that the load applied at infinity is sufficiently small so as to allow a "small strain nonlinear theory" to adequately des-cribe the field far from the crack. Then, (4.13) may be further approxi-mated to yield *the simple formula*

$$J = \frac{\pi a}{4} \frac{\sigma_\infty^2}{\{\tau'(k_\infty)\mu(k_\infty)\}^{1/2}} , \tag{4.15}$$

with $k_\infty \simeq 2(\lambda_\infty-1) \equiv 2\varepsilon_\infty$, provided $\varepsilon_\infty^2 \mu(k_\infty)/\tau'(k_\infty) \ll 1$. Consider now the small strain description of a nonlinearly elastic pure power-law material

$$\frac{\varepsilon_{ij}}{\varepsilon_0} = \frac{3}{2} \alpha \left(\frac{\sigma_e}{\sigma_0}\right)^{n-1} \frac{s_{ij}}{\sigma_0} , \qquad \varepsilon_{ij} = \frac{1}{2}(u_{i,j}+u_{j,i}) , \tag{4.16}$$

He and Hutchinson [18]. Here s_{ij} is the stress deviator and σ_e is the effective stress $\sigma_e = \sqrt{(3 s_{ij}s_{ij}/2)}$. In this case (4.15) reduces to

$$J = \pi a \sigma_\infty \varepsilon_\infty \sqrt{n} , \tag{4.17}$$

where $\varepsilon_\infty = \lambda_\infty-1$. This agrees with the result obtained in [18]. Figure 5 displays the variation of J with n according to (4.17). The finite ele-ment solution shown there is taken from Table 3 of He and Hutchinson [18]. Again, the agreement is striking.

ACKNOWLEDGEMENT

The results reported in this paper were obtained in the course of research sponsored by the U.S. National Science Foundation under Grant CME 81-06581. We gratefully acknowledge this support.

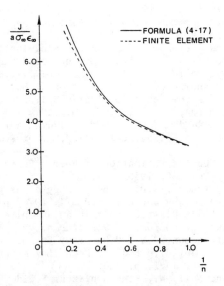

Figure 5 - Variation of the Energy Release Rate J with the Material Hardening Exponent n in Plane Strain Mode I. Comparison of Formula (4.17) with Finite Element Results [18].

REFERENCES

1. Hutchinson, J.W., *Nonlinear Fracture Mechanics*, Department of Solid Mechanics, Technical University of Denmark, 1979.
2. Rice, J.R., "Mathematical Analysis in the Mechanics of Fracture", *Fracture: An Advanced Treatise*, Editor: H. Liebowitz, Vol. 2, Academic Press, New York, 1968.
3. Stephenson, R.A., "The Equilibrium Field Near the Tip of a Crack for Finite Plane Strain of Incompressible Elastic Materials", *Journal of Elasticity*, Vol. 12, 1982, p. 65.
4. Knowles, J.K., "A Nonlinear Effect in Mode II Crack Problems", *Eng. Fracture Mechanics*, Vol. 15, 1981, p. 469.
5. Knowles, J.K., "On Finite Anti-Plane Shear for Incompressible Elastic Materials", *J. Australian Mathematical Soc.*, Series B, Vol. 19, 1977, p. 400.
6. Williams, M.L., "The Stresses around a Fault or Crack in Dissimilar Media", *Bulletin of the Seismological Soc. of America*, Vol. 49, 1959, p. 199.
7. England, A.H., "A Crack Between Dissimilar Media", *J. of Applied Mechanics*, Vol. 32, 1965, p. 400.
8. Comninou, M., "The Interface Crack", *J. of Applied Mechanics*, Vol. 44, 1977, p. 631.
9. Achenbach, J.D., Keer, L.M., Khetan, R.P. and Chen, S.H., "Loss of Adhesion at the Tip of an Interface Crack", *J. of Elasticity*, Vol. 9, 1979, p. 397.
10. Knowles, J.K. and Sternberg, E., "Large Deformations near the Tips of an Interface-Crack Between Two Neo-Hookean Sheets", *J. of Elasticity*, Vol. 13, 1983, p. 257.

11.   John, F., "Plane Strain Problems for a Perfectly Elastic Material of Harmonic Type", *Communications on Pure and Appl. Math.*, Vol. XIII, 1960, p. 239.

12.   Knowles, J.K. and Sternberg, E., "On the Singularity Induced by Certain Mixed Boundary Conditions in Linearized and Nonlinear Elastostatics", *Int. J. Solids and Structures*, Vol. 11, 1975, p. 1173.

13.   Ogden, R.W. and Isherwood, D.A., "Solution of Some Finite Plane Strain Problems for Compressible Elastic Solids", *Quart. J. of Mech. and Appl. Mathematics*, Vol. 31, 1978, p. 219.

14.   Varley, E. and Cumberbatch, E., "Finite Deformations of Elastic Materials Surrounding Cylindrical Holes", *J. of Elasticity*, Vol. 10, 1980, p. 341.

15.   Abeyaratne, R., "On the Estimation of Energy Release Rates", *J. of Appl. Mechanics*, Vol. 50, 1983, p. 19.

16.   Begley, J.A. and Landes, J.D., "The J-Integral as a Fracture Criterion", *Fracture Toughness*, ASTM STP 514, 1972, p. 1.

17.   Amazigo, J.C., "Fully Plastic Crack in an Infinite Body under Anti-Plane Shear", *Int. J. Solids and Structures*, Vol. 10, 1974, p. 1003.

18.   He, M.Y. and Hutchinson, J.W., "The Penny-Shaped Crack and the Plane Strain Crack in an Infinite Body of Power-Law Material", *J. of Appl. Mech.*, Vol. 48, 1981, p. 830.

19.   Truesdell, C. and Noll, W., "The Nonlinear Field Theories of Mechanics", *Handbuch der Physik*, Vol. III/3, Springer, Berlin, 1965.

20.   Abeyaratne, R., "Discontinuous Deformation Gradients in Plane Finite Elastostatics of Incompressible Materials", *J. of Elasticity*, Vol. 10, 1980, p. 255.

21.   Knowles, J.K., "The Finite Anti-Plane Shear Field near the Tip of a Crack for a Class of Incompressible Elastic Solids", *Int. J. of Fracture*, Vol. 13, 1977, p. 611.

22.   England, A.H., *Complex Variable Methods in Elasticity*, Wiley, London, 1971.

23.   Rice, J.R. and Sih, G.C., "Plane Problems of Cracks in Dissimilar Media", *J. of Appl. Mechanics*, Vol. 32, 1965, p. 418.

MODELLING PROBLEMS IN CRACK TIP MECHANICS
CFC10, University of Waterloo

Waterloo, Ontario, Canada
August 24-26, 1983

ADVANCED EXPERIMENTAL TECHNIQUES IN CRACK TIP MECHANICS

A.S. KOBAYASHI
Department of Mechanical Engineering
University of Washington
Seattle, Washington, U.S.A.

1. INTRODUCTION

The experimental techniques for crack tip mechanics of the 1970's were governed by the practical requirements for determining accurately 2- and 3-D stress intensity factors in linear elastic fracture mechanics (LEFM). The extensive applications of three-dimensional frozen-stress photoelasticity [1], interferometry [2] and moire method [3] yielded static stress intensity factors for complex boundary value problems, such as a corner flaw at a nozzle-cylinder junction and at a through hole [4] and a compact specimen [5]. Dynamic stress intensity factors determined by the extensive use of dynamic photoelasticity [6,7] and dynamic caustics [8,9] provided considerable insight to the controversial criteria for dynamic fracture and crack arrest. Dynamic photoelasticity and dynamic caustics were also used to establish dynamic crack curving and branching criteria [10,11], which are also applicable to static and quasi-static crack problems, and to repudiate the proposed fracture mechanics interpretation of the V-notched Charpy data [12]. These experimental techniques, which are constantly being improved to determine well-defined physical quantities, i.e., the stress intensity factors, have contributed to the credibility which linear elastic fracture mechanics commands, in postmortem failure analysis and life-time prediction of structural components.

The inevitable extensions of linear elastic fracture mechanics to fracture of composites, fatigue crack extension and stable crack growth as well as ductile fracture have imposed a new role onto the above experimental techniques. The experimental results are now also used to identify the physical laws and associated physical parameters governing these fracture phenomena. The search for these unknown physical parameters requires increased experimental accuracy as well as advanced data processing technique in the presence of geometric and material nonlinearities encountered in nonlinear fracture mechanics.

The purpose of this paper is to review the advances made in the esta-blished experimental techniques as well as to report on new experimental techniques for analyzing the traditional as well as new problems in crack tip mechanics. The techniques are discussed under three categories of 2- and 3-D linear elastic, 2-D elasto-plastic and 2-D dynamic fracture mechanics.

2. 2-D LINEAR ELASTIC FRACTURE MECHANICS

2.1 Acousto-Elasticity

Acousto-elasticity, which was hailed as an analog to photoelasticity for opaque materials in 1959 [13], failed to achieve wide acceptance due to the unresolved transducer coupling effect and high sonic attenuation [14]. The resurgence of acousto-elasticity in the 1980's is due in part to improvements in the instrumentation techniques but is mainly attributed to the ability for processing large amounts of ultrasonic data by a computer controlled scanning system [15]. Since longitudinal ultrasonic waves provide information only on plane-stress isopachics (sum of principal stresses) the use of shear waves measurements, which are referred to as acoustic anisotropy (texture) in the material can be modelled by orthotropic elasticity theory which involves three acousto-elastic constants [16,17]. The acoustic birefringence equation becomes

$$B = \{[B_0 + M_1(\sigma_1 + \sigma_2) + M_2(\sigma_1 - \sigma_2)\cos 2\theta]^2 + [M_3(\sigma_1 - \sigma_2)\sin 2\theta]^2\}^{1/2} , \qquad (1)$$

where B_0 is the initial birefringence of the unstressed state. M_1, M_2 and M_3 are the three acousto-elastic constants. θ is the angle between an acousto-elastic axis and a principal stress direction.

The angle between the initial and stressed acousto-elastic axes, which in general do not coincide with the principal stress axes, is

$$\tan 2\phi = \frac{M_3(\sigma_1 - \sigma_2)\sin 2\theta}{B_0 + M_1(\sigma_1 + \sigma_2) + M_2(\sigma_1 - \sigma_2)\cos 2\theta} . \qquad (2)$$

The shear stress in the xy plane is then given by

$$\sigma_{xy} = \frac{B \sin 2\phi}{2M_3} \qquad (3)$$

Clark, Mignogna and Sanford [18] used the above relations to measure the stress intensity factor in a 2024-T351 aluminum compact specimen shown in Figure 1. A pulse-echo-overlap system, as shown in Figure 2, was used to determine point-by-point, the orientation of the acoustic axes and the

acoustic birefringence in the 51 x 51 mm square region shown in Figure 1.
A 10 MHz ac-cut quartz shear-wave transducer of 1.8 mm diameter was used
in a manual scanning process. The estimated accuracy were approximately
5 percent and ± 2 degrees in birefringence and ϕ measurements, respectively.

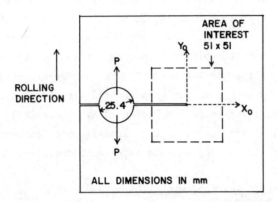

Figure 1 - Schematic Diagram of the Modified Compact Tensile Specimen.
The Region where Acoustic Measurements were made is Labelled
'Area of Interest'. Also shown are the initial acoustic axes,
X_0 and Y_0.

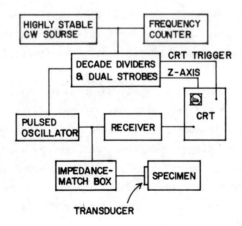

Figure 2 - Block Diagram of Pulse-Echo-Overlap System

The acousto-elastic birefringence generated from 66 data points in the
square region was reconstructed and Sanford's procedure [19] was used to
compute five coefficients in the LEFM crack tip stress field by averaging

the results of 100 computations using 20 randomly selected data points
each time. Good agreement between the corresponding coefficients, which
were obtained from a similar photoelasticity experiment, were noted. The
stress intensity factor was computed from the coefficient of the first
term, or the $1/\sqrt{r}$ term, in the above polynomial crack tip stress field.

The acousto-elastic technique is one of the few static, stress analysis
techniques available for opaque materials. As in 2-D photoelasticity, the
thickness-averaged acoustic birefringence is not subject to the plane
stress constraint of the caustic method. Obvious improvement in the technique can be made by incorporating an automated scanning procedure with
real-time data processing which has been used by others [15]. Yet to be
explored is the physical significance of acoustic birefringence associated
with the crack tip plastic region associated with ductile fracture.

3. 3-D LINEAR ELASTIC FRACTURE MECHANICS

3.1 Frozen Stress-Moire Technique

The hybrid technique, which utilizes both frozen stress, 3-D photoelasticity and moire interferometry, provides the complete information for
characterizing the crack tip state [20]. The procedure is redundant in
that the in-plane displacement field, which is determined by the high resolution moire technique [21], also defined the strain and stress fields.
The isochromatics, however, can be used to verify the accuracy of the
stresses which are obtained by numerically differentiating the displacements. Such optimum use of the redundant experimental data is yet to be
explored.

The procedure consists of applying an aluminum reflective grating to
the slices cut from the frozen-stress 3-D photoelastic model and returning
the slice to its unloaded stage by annealing through its critical temperature. The in-plane displacements are obtained by moire interferometry of
the deformed grating superimposed onto an undeformed virtual grating with
a grating density of 2400 lines per mm. Figure 3 shows the experimental set
for viewing the Moire fringes. The in-plane displacements of u_n and u_z
are related to the stress intensity factor by:

for plane strain;

$$u_n = \frac{K_{AP}}{G} \sqrt{\frac{r}{2\pi}} \cos \frac{\theta}{2} \left[1 - 2\nu + \sin^2 \frac{\theta}{2}\right] , \qquad u_z = \frac{K_{AP}}{G} \sqrt{\frac{r}{2\pi}} \sin \frac{\theta}{2} \left[2 - 2\nu\cos^2 \frac{\theta}{2}\right] ,$$

$$(4)$$

for plane stress;

$$u_n \doteq \frac{K_{AP}}{G} \sqrt{\frac{r}{2\pi}} \sin \frac{\theta}{2} \left[\frac{1-\nu}{1+\nu} + \sin^2 \frac{\theta}{2} \right] , \qquad u_z = \frac{K_{AP}}{G} \sqrt{\frac{r}{2\pi}} \sin \frac{\theta}{2} \left[\frac{2}{1+\nu} - \cos^2 \frac{\theta}{2} \right] ,$$

$$(5)$$

where G is the shear modulus of elasticity and ν is Poisson's ratio.

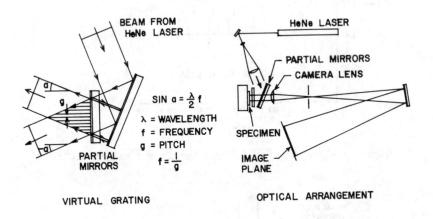

Figure 3 - Setup for Moire Interferometry

The photoelastic-moire technique was used to determine the variation in stress intensity factor along a straight crack front in a four-point bend specimen of 279.4 x 25.7 x 13.3 mm size after ASTM E399. Figure 4 shows the stress intensity factors at the centre slice of the cracked beam determined by both photoelasticity and moire interferometry for a crack depth to beam ratio of 0.5. The reference K_{th} in Figure 4 was determined by 2-D plane strain analysis [22]. Figure 5 shows the variation of stress intensity factor through the thickness of the beam. A state of plane stress and the presence of a $1/\sqrt{r}$ singularity were assumed in the data reduction process.

While the uncertainties in the relaxation mechanism as well as the resultant state of stress associated with the annealing process require further studies, the frozen stress-moire techniques provides a mean for complete and detailed stress analysis of the crack tip state in 3-D linear elastic fracture mechanics.

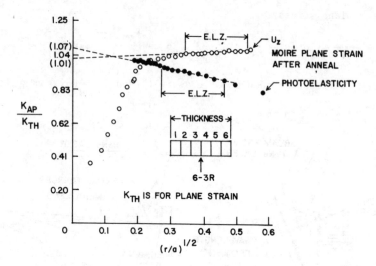

Figure 4 - Comparison of Photoelastic and Moire Results

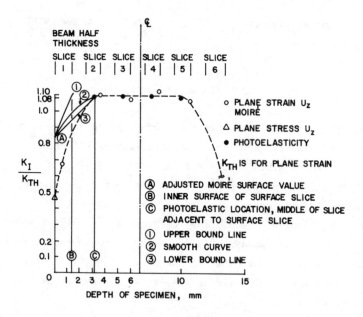

Figure 5 - SIF Distribution Across the Thickness
          (a/W = 0.50)

3.2 Isodyne Photoelasticity

Isodyne represents curves of constant intensity of the normal forces acting on the characteristic curves in a plane stress field and are thus related to the first derivatives of the Airy stress function. Two isodyne fields related to two orthogonal characteristic curves completely define the elastic state of plane stress [23]. When modelled optically with the integrated polariscope, shown in Figure 6 [24], the photoelastic isodynes resemble the isochromatics generated by scattered light photoelasticity. Similar to scattered light photoelasticity, optical inhomogeneity generated by the high stress gradient in the vicinity of the crack tip may distort the photoelastic isodyne. The requirement for a plane stress state, which is not a prerequisite in scattered light photoelasticity, can be modulated by the "semi-plane stress state" used by Pindera et al [25] who then determined the stress intensity factor at the midsection, i.e., plane of symmetry, of a four-point bend specimen shown in Figure 7. Also shown in Figure 7 is the variation in the stress intensity factor computed for various crack tip distance where a pronounced effect of the near-tip nonlinearity and crack tip bluntness are noted.

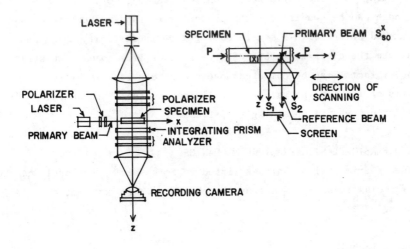

Figure 6 - Integrated Polariscope for Isodyne Photoelasticity

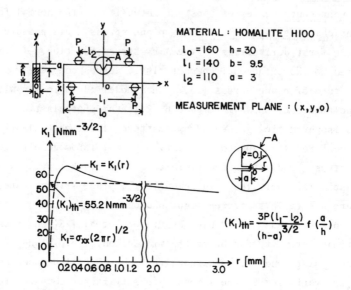

Figure 7 - Four Point Bend Beam with Sharp Notch. $K_I(r)$ for Central Plane was Determined by Isodyne Technique.

Assuming that the influence of optical inhomogeneity in the scattered light path can be quantified, the photoelastic isodyne technique share the same advantage of 3-D scattered light photoelasticity which can be used to analyze the crack tip state of stress under live load. The stress intensity factor can be computed more accurately if K is expressed directly in terms of the isodyne value thus eliminating the extra numerical differentiation process in obtaining the stresses.

4.   2-D ELASTO-PLASTIC FRACTURE MECHANICS

The experimental techniques listed in this section obviously can be used for elastic analysis but unlike the above, are not limited to elastic analysis.

4.1 Moire Technique

The use of moire technique in elasto-plastic fracture mechanics is not new [3,26]. Despite its obvious application to high temperature, nonlinear problems in fracture mechanics, literature is relatively sparse in the fracture mechanics interpretation of the crack tip displacement field

determined by the moire method. Exception to the above is the analysis
of externally notched rings sliced from a Type 304 stainless tube,
7.1 mm O.D. and 0.38 mm thick, with electro-etched cross-line gratings
of 40 lines per mm and subjected to a simulated internal pressure at
1100 F [27]. Figure 8 shows the experimental setup for recording the
distorted grating which was analyzed by master gratings of 4 and 8 lines/mm.
From the resultant u and v moire fringe patterns, COD for slow-crack growth
initiation was found to be

$$COD = 0.976 \cdot a \cdot \sigma^{5.78} , \qquad (6)$$

where the crack length a and the applied hoop stress ϕ are represented in
terms of mm and KN, respectively.

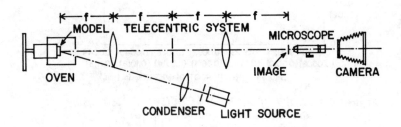

Figure 8 - Optical Setup for High-Temperature Studies

Figure 9 shows that the initiation COD in this experiment remained
relatively constant despite the changes in the crack tip bluntness.
Sciammarella then estimated the J-integral for the initiation of slow
crack growth by the following approximate formula after Rice et al [28]

$$J_{in} = \frac{1}{bt} \left[2 \int_0^{\delta_{cr}} Pd\delta_{cr} - P \cdot \delta_{cr} \right] , \qquad (7)$$

where b is the ligament length, t is the specimen thickness, δ_{cr} is the
displacement due to the presence of the crack between two reference sections for the load at the moment of crack initiation and

$$P = \sigma A , \qquad (8)$$

where σ is the hoop stress and A is the specimen cross-sectional area. The
values of δ_{cr} were obtained as

$$\delta_{cr} = \delta_{total} - \delta_{nocr} \; , \tag{9}$$

where δ_{total} is the displacement between two reference cross sections and δ_{nocr} is the displacement given by

$$\delta_{nocr} = \Delta\varepsilon_h \; . \tag{10}$$

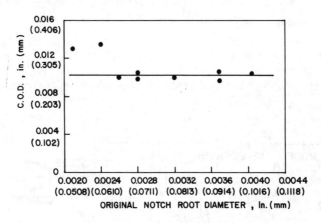

Figure 9 - *C.O.D. for the Initiation of Slow Crack Growth as a Function of Original Notch-Root Diameter*

Table 1 - J_{in} *for Ring Specimen at 1100°F*

Ring No.	$P \cdot \delta$, KN·cm	J_{in} MJ/m^2
1	0.0682	0.03980
2	0.0565	0.05554
3	0.0455	0.04755
4	0.0661	0.04520
5	0.0517	0.04762
6	0.0298	0.02206
7	0.0684	0.04681
8	0.0421	0.03484
9	0.0709	0.04742

Table 1 shows the excess variations in the J estimated by this procedure thus leading this author to conclude that COD is a better criterion for predicting the initiation of slow crack growth.

Moire method, which was limited in its applications to fracture problems involving large scale yielding due to its low sensitivity, can be
used in the high sensitivity region of linear elastic fracture mechanics
by the recent developments in high density line gratings up to 4000 lines
per mm with grating sizes up to 100 x 63 mm [29]. The use of virtual
grating, which was described previously, eliminates the need for physical
contact of the reference grating. Its use at elevated temperature testing,
such as that described above, or under an explosive loading condition may
be in doubt since the long optical paths, which is required in the experimental setup, may be distorted by the moving air current or shock waves.

The moire fringes can be generated by holographic interferometry.
Referred to as "intrinsic holographic moire", these fringes can be recorded
by using the basic setup shown in Figure 10 [30]. The reference state is
obtained by a single exposure of the unloaded specimen. Rigid body motions
of the loaded specimen are compensated by displacing the reference state
and observing the fringe contrast in the TV monitor. The u and v fringe
patterns are recorded on tape or alternatively photographed directly.

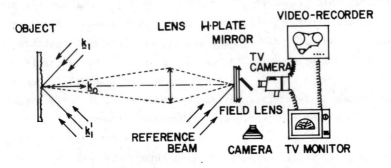

*Figure 10 - Schematic Representation of the Recording
System in Image Plane Holography*

4.2 Laser Speckle Method

Despite its many implied applications in fracture mechanics [31,32],
literature is void of useful data which has been generated by the speckle
method. With its high sensitivity, i.e., u and v displacement measurements
of the order of 0.005 mm, the laser speckle method should find wide ranging
applications in experimental fracture mechanics. By using the digital
imaging technique [33,34] to cross correlate the two speckle images gen-

erated by the unloaded and loaded specimens, the method provides an efficient procedure for processing the immense amount of data and for easy
access to graphic peripherals.

4.3 Hybrid Experimental-Numerical Analysis

One of the major obstacles, which hinders the progress of experimental
ductile fracture research, is the undefined crack-tip states of stress and
strain in the presence of large scale yielding. Since the $1/\sqrt{r}$ singular
state in linear elastic fracture mechanics is a physical impossibility
which successfully models brittle fracture, similar phenomenological
models could be developed for a crack under large scale yielding. A popular
and possibly over-exploited model is the Dugdale strip yield zone which
conveniently reduces the elastic-plastic crack-tip state to an elastic
one. The Dugdale strip yield model used in a recent analysis [37] is a
modification of the classical Dugdale model where higher order terms were
added to increase the number of disposable parameters. Experimental data
is then used to fit the disposal parameters associated with the Dugdale
model, which is modified to fit the complex state associated with large
scale yielding. This procedure is the plastic analog of the elastic procedure
of determining the stress intensity factor from photoelasticity and moire
fringe data. The adequacy of such a model can be verified by matching other
crack-tip data which is not used in the fitting process but which is generated numerically by the Dugdale model and independently by the experiment.
The extensive numerical experimentation necessary for this verification study
in essence replaces the finite element or boundary element method used in
the traditional hybrid experimental-numerical stress analysis technique [36].
The modified Dugdale model which is verified through the generation mode of
hybrid experimental-numerical analysis, can then be used to generate numerical
various fracture parameters for evaluation.

The utility of the hybrid experimental-numerical analysis is demonstrated
by a recent investigation on stable-crack growth under mixed-mode loading
[37]. Isochromatics in a 1.6 mm thick polycarbonate tensile specimen with
slanted central crack (SCN) were recorded during a continuing stable
crack growth period. The resultant Z-shaped crack was modelled by a
straight Dugdale crack, which was modified to account for the residual

stresses left behind in the wake of the rapidly extending crack, as shown
in Figure 11. The modification consisted of two unknown tangential forces
acting at the physical crack tip. Lengths of the Dugdale strip yield zones
ahead of the crack tip were measured from the photoelastic records [37].
These lengths coincided with the length of the theoretical values of the
horizontal crack thus justifying the use of the model in Figure 11 to
represent the Z-shaped cracks. The crack-tip stress field which is represented by a polynomial stress function of the crack-tip coordinates together
with the two unknown tangential forces were fitted to the recorded elastic isochromatics surrounding the plastic region using an overdeterministic
fitting routine [38]. Figure 12 shows the near- and far-field isochromatics
which were regenerated by using the modified Dugdale model and those
obtained by photoelasticity. Figure 13 shows the crack tip opening angle
(CTOA), which was computed by using the modified Dugdale model, for the
two initial crack geometries to be almost constant during the stable crack
growth process.

  While the hybrid experimental-numerical technique may not provide the
micromechanics insight to crack-tip mechanics, it can be used to effectively
extract fracture parameters which otherwise cannot be measured directly.

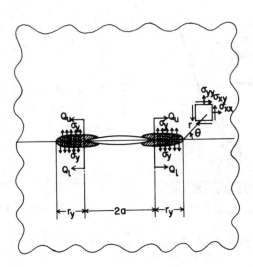

Figure 11 - Modified Dugdale Strip Yield Model

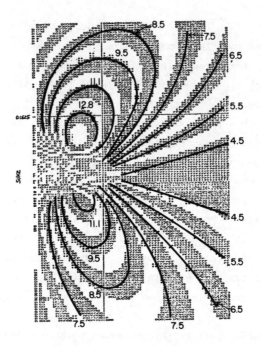

Figure 12 - *Computer Generated and Actual (Solid
Curves) Isochromatics 30° SCN Specimen*

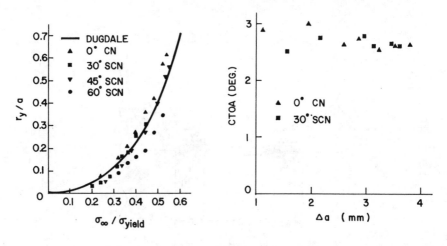

Figure 13 - *CTOA during Stable Crack Growth of
0° CN and 30° SCN*

4.4 Caustic Method

The method of caustics has become a popular technique for measuring the static and dynamic stress intensity factors for plane-stress problems in linear elastic fracture mechanics. Caustic can also be generated by any deformed specimen surface including the obvious dimpling surrounding a ductile crack. Rosakis and Freund [39] used an asymptotic elastic-plastic analysis to relate this dimpling to a plastic intensity factor. By postulating an HRR singularity, J-deformation theory of plasticity and the separation of θ and r, the plastic strain in the thickness direction is obtained as

$$\varepsilon^p_{33} = -(\varepsilon^p_{rr} + \varepsilon^p_{\theta\theta}) \, , \tag{11}$$

where the in-plane plastic strain components are given in terms of the stress components as

$$e_{rr} = \frac{\alpha\sigma_0}{E} \left[\frac{JE}{\alpha\sigma_0^2 I_n r} \right]^{\frac{n}{n+1}} \left(\frac{\sigma_e}{\sigma_0} \right)^{n-1} \left(\Sigma_{rr} - \frac{1}{2} \Sigma_{\theta\theta} \right) \, , \tag{12}$$

$$e^p_{\theta\theta} = \frac{\alpha\sigma_0}{E} \left[\frac{JE}{\alpha\sigma_0^2 I_n r} \right]^{\frac{n}{n+1}} \left(\frac{\sigma_e}{\sigma_0} \right)^{n-1} \left(\Sigma_{\theta\theta} - \frac{1}{2} \Sigma_{rr} \right) \, .$$

The resultant caustic, of height D, generated by the thickness direction strain of equation (12) is shown in Figure 14. J-integral value can then be determined by

$$J = \frac{\sigma_0 D}{13.5 z_0 d} \, , \tag{13}$$

where z_0, d and σ_0 are the screen distance, specimen thickness and tensile yield stress, respectively.

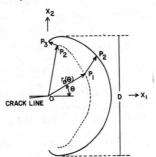

Figure 14 - Geometric Construction of the Predicted
Initial Curve (dashed) and Caustic Curve (solid)

While further verification study is necessary, the caustic method promises to provide an experimental procedure with which, the J-value can be determined directly using crack tip measurements in contrast to the ASTM designated far-field procedure whish is based on many simplifying assumptions.

5. 2-D DYNAMIC FRACTURE MECHANICS

As mentioned in the Introduction, literature is abundant with experimental results on linear elastic dynamic fracture using dynamic photoelasticity and dynamic caustics. Experimental as well as data processing procedures for these two techniques are continually being improved and their domain of application is being extended. One such extension is the use of the hybrid experimental-model analysis for modelling the Dugdale strip yield zone ahead of a rapidly tearing crack [37]. Likewise, the caustic method with its asymptotic elastic-plastic solution could be extended with relative ease to analyze problems involving rapid tearing.

6. CLOSING COMMENTS

While no claim is made for completeness, most of the significant new experimental techniques for crack tip mechanics hopefully have been mentioned in this paper. The potential of applying some of the 2-D techniques, which were listed under specific fields in crack tip mechanics, to other fields obviously must be explored.

ACKNOWLEDGEMENT

The work reported here was obtained under ONR Contract N00014-76-C-0060 NR 064-478. The authors wish to acknowledge the support and encouragement of Dr. Y. Rajapakse, Office of Naval Research, during the course of this investigation.

REFERENCES

1. SMITH, C.W., "Use of Three-Dimensional Photoelasticity and Progress in Related Areas", *Experimental Techniques in Fracture Mechanics*, Vol. 2, edited by A.S. Kobayashi, Society for Experimental Stress Analysis, 1975, pp. 3-58.
2. PACKMAN, P.F., "The Role of Interferometry in Fracture Studies", *Experimental Techniques in Fracture Mechanics*, Vol. 2, edited by A.S. Kobayash, Society for Experimental Stress Analysis, 1975, pp. 59-87.

3. LIU, H.W. and KE, J.S., "Moire Method", *Experimental Techniques in Fracture Mechanics*, Vol. 2, edited by A.S. Kobayashi, Society for Experimental Stress Analysis, 1975, pp. 111-165.

4. SMITH, C.W., POST, D. and NICOLETTO, G., "Prediction of Sub-critical Crack Growth from Model Experiments", *Developments in Theoretical and Applied Mechanics*, edited by T.J. Chung and G.R. Karr, The University of Alabama in Huntsville, 1982, pp. 167-179.

5. FOURNEY, M.E., "Experimental Determination of the Effect of Crack Front Curvature in an ASTM Compact Tension Specimen", *Proc. of the Fourth Brazilian Congress of Mechanical Engineering*, 1977, pp. 13-26.

6. DALLY, J.W., "Dynamic Photoelastic Studies of Fracture", *Experimental Mechanics*, Vol. 19, 1979, pp. 349-361.

7. BRADLEY, W.B. and KOBAYASHI, A.S., "Fracture Dynamics - A Photoelastic Investigation", *J. of Eng. Fracture Mechanics*, Vol. 3, 1971, pp. 317-332.

8. KALTHOFF, J.F., BEINERT, J. and WINKLER, S., "Measurements of Dynamic Stress Intensity Factors for Fast Running and Arresting Cracks in Double Cantilever-Beam Specimens", *Fast Fracture and Crack Arrest*, edited by G.T. Hahn and M.F. Kanninen, ASTM STP 627, 1977, pp. 616-176.

9. THEOCARIS, P.S., "Optical Method of Caustics in the Study of Dynamic Problems of Running Cracks", *Optical Engineering*, Vol. 21, 1982, pp. 581-601.

10. RAMULU, M. and KOBAYASHI, A.S., "Dynamic Crack Curving - A Photoelastic Evaluation", *Experimental Mechanics*, Vol. 22, No. 3, 1983, pp. 1-9.

11. RAMULU, M., KOBAYASHI, A.S. and KANG, B.S-J., "Dynamic Crack Branching - A Photoelastic Evaluations", to be published, *ASTM-STP*.

12. KALTHOFF, J.F., WINKLER, S., BOHME, W. and KLEMM, W., "Measurements of Dynamic Streee Intensity Factors in Impacted Bend Specimens", *C.S.N.I. Specialist Meeting on Instrumented Precracked Charpy Testing*, edited by R.A. Wullaert, Electric Power Research Institute, EPRI NP-2102-LD, 1981, pp. 2-3 - 2-19.

13. BENSON, R.W. and RAELSON, V.J., "Acoustoelasticity", *Product Engineering*, Vol. 30, 1959, pp. 56-59.

14. BLINKA, J. and SACHSE, "Application of Ultrasonic-Pulse-Spectroscopy Measurements to Experimental Stress Analysis", *Experimental Mechanics*, Vol. 16, 1976, pp. 448-453.

15. KINO, G.S., HUNTER, J.B., JOHNSON, G.C., SELFRIDGE, A.R., BARNETT, D.M., HERMAN, G. and STEELE, C.R., "Acoustoelastic Imaging of Stress Fields", *J. of Applied Physics*, Vol. 50, No. 4, 1979, pp. 530-535.

16. OKADA, K., "Stress-Acoustic Relation for Stress Measurements by Ultrasonic Technique", *J. of Acoustical Soc. of Japan (E)*, Vol. 1, No. 3, 1980, pp. 193-200.

17. OKADA, K., "Acoustoelastic Determination of Stress in Slightly Orthotropic Materials", *Experimental Mechanics*, Vol. 21, 1981, pp. 461-466.

18. CLARK, A.V., MIGNOGNA, R.B. and SANFORD, R.J., "Acousto-Elastic Measurement of Stress and Stress Intensity Factors Around Crack Tips", *Ultrasonics*, March, 1983, pp. 57-64.

19. SANFORD, R.J. and CHONA, R., "An Analysis of Photoelastic Fracture Patterns with Sampled Least-Squares Methods", *Proc. of the 1981 SESA Spring Meeting*, SESA, June, 1981, pp. 273-276.

20. SMITH, C.W., POST, D. and NICOLETTO, G., "Experimental Stress Intensity Distributions in Three Dimensional Cracked Body Problems", *Proc. of the Joint Conference on Experimental Mechanics*, SESA, May, 1982, pp. 196-200.

✗ 21. NICOLETTO, G., POST, D. and SMITH, C.W., "Moire Interferometry for High Sensitivity Measurements in Fracture Mechanics", *Proc. of the Joint Conference on Experimental Mechanics*, SESA, May, 1982, pp. 266-270.

22. "Standard E399 on Fracture Toughness Testing", *Annual Book of ASTM Standards*, Part 10 - Metals, 1981, pp. 605-607.

23. PINDERA, J.T., "New Development in Photoelastic Studies: Isodyne and Gradient Photoelasticity", *Optical Engineering*, Vol. 21, July/August, 1982, pp. 672-678.

24. MAZURKIEWICS, S.B. and PINDERA, J.T., "Integrated Plane Photoelastic Method - Application of Photoelastic Isodynes", *Experimental Mechanics*, Vol. 19, July, 1979, pp. 225-234.

25. PINDERA, J.T., KRASNOWSKI, B.R. and PINDERA, M.-J., "An Analysis of Semi-Plane Stress States in Fracture Mechanics and Composite Structures Using Isodyne Photoelasticity", *Proc. of the 1982 Joint Conf. on Experimental Mechanics*, SESA, May, 1982, pp. 417-421.

✗ 26. HU, W.L. and LIU, H.W., "Crack Tip Strain - A Comparison of Finite Element Calculations and Moire Measurements", *Cracks and Fracture*, edited by J.L. Swedlow and M.L. Williams, ASTM STP 601, 1976, pp. 622-634.

27. SCIAMMARELLA, C.A. and RAO, M.P.K., "Failure Analysis of Stainless Steel at Elevated Temperatures", *Experimental Mechanics*, Vol. 19, 1979, pp. 389-398.

28. RICE, J.R., PARIS, P.C. and MERKLE, J.G., "Some Further Results on J-Integral Analysis and Estimates", *Progress in Flaw Growth and Fracture Toughness Testing*, ASTM STP 536, 1973, pp. 231-245.

✗ 29. POST, D., "Developments in Moire Interferometry", *Optical Engineering*, Vol. 21, May/June, 1982, pp. 458-467.

30. SCIAMMARELLA, C.A., "Holographic Moire, An Optical Tool for the Determination of Displacements, Strains, Contours and Slopes of Surfaces", *Optical Engineering*, Vol. 21, May/June, 1982, pp. 447-457.

31. CHIANG, F.P., ADACHI, J., ANASTASI, R. and BEATTY, J., "Subjective Laser Speckle Method and Its Application of Solid Mechanics Problems", *Optical Engineering*, Vol. 21, May/June, 1982, pp. 379-390.

32. BOONE, P.M., "Use of Close Range Objective Speckles for Displacement Measurement", *Optical Engineering*, Vol. 21, May/June, 1982, pp. 407-410.

33. PETERS, W.H. and RANSON, W.F., "Digital Imaging Techniques in Experimental Stress Analysis", *Optical Engineering*, Vol. 21, May/June, 1982, pp. 427-431.

34. McNEIL, S.R., PETERS, W.H., RANSON, W.F. and SUTTON, M.A., "Digital Image Processing in Fraucture Mechanics", *Proc. of the 1983 SESA Spring Meeting*, Cleveland, Ohio, May, 1983.

35. KOBAYASHI, A.S. and LEE, O.S., "Elastic Field Surrounding a Rapidly Tearing Crack", *Elastic-Plastic Fracture Mechanics*, ASTM STP, to be published.

36. KOBAYASHI, A.S., "Hybrid Experimental-Numerical Stress Analysis", *Experimental Mechanics*, Vol. 23, 1983, pp. 338-347.

37. SUN, Y.J., LEE, O.S. and KOBAYASHI, A.S., "Crack Tip Plasticity Under Mixed-Mode Loading", *Proc. of the ICF Symposium on Fracture Mechanics*, Science Press, Beijing, China, November, 1983, pp. 582-588.

38. BETSER, A.A., KOBAYASHI, A.S., LEE, O.S. and KANG, B.S.-J., "Crack Tip Dynamic Isochromatics in the Presence of Small Scale Yielding", *Experimental Mechanics*, Vol. 22, 1982, pp. 132-138.

39. ROSAKIS, A.J. and FREUND, L.B., "Optical Measurement of Plastic Strain Concentration at a Crack Tip in a Ductile Steel Plate", *ASME, J. of Eng. Materials and Tech.*, Vol. 104, 1982, pp. 115-120.

LIFETIME PREDICTION FOR METALLIC STRUCTURAL COMPONENTS SUBJECTED TO
LOADING WITH VARYING INTENSITY

H.H.E. LEIPHOLZ
Department of Civil Engineering
University of Waterloo
Waterloo, Ontario, Canada

1. INTRODUCTION

From 1852 to 1870, August Wöhler performed his pioneering experiments
on fatigue failure of railway coach axles, introduced the concept of the
S,N-curve, and determined the fatigue limit of various metallic materials.
His achievements have been outstanding and so impressive, that the concepts
and methods he created were accepted unchallenged by other researchers and
engineers. Sometimes, his theories were unfortunately applied even under
circumstances which were different from those under which his approach
and his solutions were valid. The experimental setup and the main problems
he had chosen, allowed him for example to use the stress amplitude as one
of the fatigue parameters. This choice seemed to be so obvious and straight
forward that it was retained also in cases for which it probably would not
be admissible anymore, for example, in the presence of loading with varying intensity.

In fact, in order to be able to cover general cases of loading, one
should create the motion of a more or less abstract fatigue parameter,
i.e., a kind of "fatigue norm", of which the stress amplitude would prove
to be a special case. Such procedure has to be more successful than one
by which one would try to save a simple notion, like the stress amplitude,
for the interpretation of a more sophisticated loading situation, probably
without satisfactory results.

Another point is that the early, phenomenological theory of metal
fatigue did mostly describe apparent effects only, without trying to
explain or predict their occurrence by taking the microscopic structure
of the respective metal into account. The truth is that metal fatigue
mechanics cannot exist without a close association with metallurgy. Only
on the basis of a theory strongly oriented towards physics, and a clari-
fication of the structural changes within the metal during cycling of the

load, can one expect to obtain fatigue life data which are close to reality.

Furthermore, one faces for loading with a varying intensity the problem of how to account for damage accumulation. A natural and logically appealing solution to this problem is to apply Miner's rule. One is certainly able to define this rule in the light of proper assumptions. However, when checking these assumptions, one realizes that one of them consists in considering the damage increment per load cycle as being a physical invariant. Yet, this is not true, as over a number of cycles the internal structure of the metal changes and therefore its capability to resist damage. Hence, time has to be taken into account (reflected perhaps in the form of number of cycles), and one has actually a "rheological" problem at hand.

Finally, it may be mentioned that due to the complexity of the factors involved in fatigue mechanics, i.e., metallic structure, damage growth, load history, etc., one is well advised to try setting up a stochastic theory of metal fatigue. Also then, one faces the basic problem of choosing the proper parameters for the description of the probabilities involved. Expectations and variances may not yield a sufficient accuracy for fatigue life predictions in all cases. One may have to try to work with probability densities instead.

In the paper, questions like the ones just raised are treated. Definite answers may not always be available. But the questions by themselves may already be of some value: They may indicate in which direction future research should be oriented.

2. ON DAMAGE ACCUMULATION

Assume a basic fatigue test to be carried out. It consists of a sinusoidal, pulsating load with stress amplitude σ_a, zero mean stress, and constant frequency being applied to the metallic structural component until fatigue fracture occurs. The number N of load repetitions needed to cause failure is recorded as the fatigue lifetime of the specimen.

Early phenomenological theories, used to explain the damage caused by such repeated loading, were based on some axiomatic assumptions. The most simple assumptions of that kind would lead to the following set of W (Wöhler) axioms:

W_1: There exists a well defined amount of damage D at which the metallic specimen fails by fatigue. This amount is characteristic for the kind of metal involved.

W_2: Each stress cycle of number **i**, $i = 1,2,3,...$, represents an equally damaging event causing the damage d_i. The damages d_i occurring per cycle are therefore identical, i.e.,

$$d_1 = d_2 = d_3 = \cdots = d = \text{const.}$$

W_3: The damage d_i per cycle is proportional to the stress amplitude σ_a, i.e.,

$$d_1 = d_2 = d_3 = \cdots = d = p\sigma_a , \qquad p = \text{const} . \qquad (1)$$

W_4: The total fatigue damage D is the sum of those damages per cycle, which exceed the threshold

$$d_0 = p\sigma_0 . \qquad (2)$$

Hence,

$$D = \sum_{i=1}^{N} (d_i - d_0) ,$$

and by virtue of (1) and (2),

$$D = p(\sigma_a - \sigma_0) \sum_{i=1}^{N} 1 = Np(\sigma_a - \sigma_0) . \qquad (3)$$

W_5: Due to the first application of the load with amplitude σ_a, the specimen suffers an initial damage D_0, which in correspondence to (3) may be given as

$$D_0 = N_0 p(\sigma_a - \sigma_0) , \qquad (4)$$

where N_0 is an appropriately chosen number.

Using the preceding axioms, one can conclude that the total damage occurring in the specimen due to the repeatedly applied load is

$$D + D_0 = (N + N_0) p(\sigma_a - \sigma_0) . \qquad (5)$$

With

$$k = \frac{D + D_0}{p} , \qquad (6)$$

relationship (5) can be changed into

$$(\sigma_a - \sigma_0) = \frac{k}{N+N_0} \; . \tag{7}$$

Setting

$$\sigma_a - \sigma_0 = \sigma^* \; , \qquad N+N_0 = N^* \; , \tag{8}$$

one has

$$\sigma^* = \frac{k}{N^*} \; , \tag{9}$$

which is a hyperbola. Therefore, the conclusion is, that, if axioms W_1 to W_5 would hold true, the experimentally derived Wöhlercurve, obtained by plotting stress amplitudes σ_a versus fatigue lifes N, had to be a shifted hyperbola, see Figure 1. Obviously, σ_B is the *ultimate strength* of the specimen at which it breaks without cycling; i.e., at N = 0. Moreover, σ_0 is the *fatigue limit*, as for stress amplitudes below σ_0, fatigue fracture is excluded.

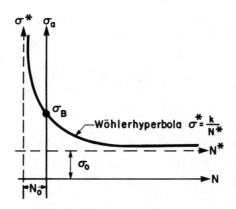

Figure 1 - Idealized Wöhlerhyperbola

As already shown by Weibull, experimental evidence leads to the conclusion that Wöhlercurves in the σ_a,N - plane do not consist of shifted hyperbolas, but are rather curves described by an equation of the type

$$(\sigma^*)^\beta = \frac{K}{N^*} \; , \qquad \beta > 1 \; . \tag{10}$$

Comparing (10) with (9), one realizes that due to the occurrence of exponent β in (10), the hyperbola in [9] had suffered stretching while being transformed into the new curve. This stretching of curve (9) indicates that there is a discrepancy in the previously assumed correlation between the damage characteristic σ_a and the fatigue life N. The reason for this discrepancy may be the fact, that phenomenologically equal events, like load cycles, may actually not be equally damaging events at all, in contradiction to W_2. This may come about through "aging" of the metal under repeated loads so that the same load cycle, if performed after a whole sequence of identical cycles, may cause more or less damage than if it were applied to the yet unloaded metal. Hence, there is a dependence of cycle damage on time. Obviously, a "rheological" effect is involved.

Under these circumstances, one must also conclude, that axiom W_3 cannot be true. In W_3, it is for example implied, that the stress amplitude is indicative for the damage occurring during a cycle. This assumption may have to be dropped in favour of a more appropriate damage characteristic. This has indeed been done. For example, Manson [2] and Coffin [3] have proposed the plastic-strain amplitude as such a characteristic, and Smith, Watson, and Topper [4] have chosen an energy-like product of stress and strain for a characterization of fatigue damage.

In the light of the preceding considerations, let a set of modified axioms for the damage assessment under repeated loading be proposed.

L_1: There exists a well defined amount of damage D at which the metallic specimen fails by fatigue. This amount is characteristic for the kind of metal involved. (This axiom is identical with W_1.)

L_2: A set of equally damaging events e_i, can be identified in the context of the load history. The damages d_i for any of the individual events e_i are identical, i.e.,

$$d_1 = d_2 = d_3 = \cdots = d = \text{const} .$$

L_3: The damage d_i per event e_i is a function of a properly chosen damage characteristic c, i.e.,

$$d_i \equiv d = f(c) \qquad \text{for any i .} \tag{11}$$

L_4: The total fatigue damage D is the sum of those damages per event e_i, which exceed the threshold

$$d_0 = f(c_0) . \tag{12}$$

Hence,

$$D = \sum_{i=1}^{N} (d_i - d_0) ,$$

which by virtue of (11) and (12) becomes

$$D = \sum_{i=1}^{N} [f(c) - f(c_0)] = [f(c) - f(c_0)] \sum_{i=1}^{N} 1 ,$$

$$D = [f(c) - f(c_0)]N . \tag{13}$$

L_5: Due to the first application of the event e_i the specimen suffers an initial damage D_0, which in correspondence to (13) may be assumed as:

$$D = [f(c) - f(c_0)]N_0 , \tag{14}$$

where N_0 is an appropriately chosen number.

On the basis of axioms L_1 to L_5, the total damage occurring in the specimen can be assessed by the expression

$$D + D_0 = (N + N_0)[f(c) - f(c_0)] ,$$

from which follows

$$f(c) - f(c_0) = \frac{D + D_0}{N + N_0} . \tag{15}$$

With

$$f(c) - f(c_0) = y - y_0 , \qquad N + N_0 = x + x_0 , \qquad D + D_0 = k ,$$

relationship (15) yields

$$y - y_0 = \frac{k}{x + x_0} , \tag{16}$$

which is nothing more than the Wöhlerhyperbola shown in Figure 1. This fact comes even clearer to light by setting

$$y - y_0 = y^* , \qquad x + x_0 = x^* ,$$

which transforms (16) into

$$y^* = \frac{k}{x^*} .$$

This equation is indeed analogous to (9).

The conclusion to be drawn at this point is, that, if it were possible to identify *physically identical* events, i.e., equally damaging events, the Wöhlerhyperbolic would materialize. Unfortunately, one may not be able to easily count physically identical events. Instead, one may have to count *phenomenologically equal* events like cycles. Then, the preceding simple axioms have still to be replaced by other ones, which take the aging of the metal into account. This has to be done, even if a more appropriate damage characteristic than the stress amplitude σ_a had been chosen.

Let now those assumptions be listed which have to be made if one chooses to work with phenomenologically equal events:

L_1^*: Identical with L_1.

L_2^*: A set of phenomenological events ε_i can be identified in the context of the load history. To each event ε_i there corresponds a distinct amount of damage d_i.

L_3^*: The damage d_i per event ε_i is a function of a properly chosen damage characteristic c and of the damage accumulated up to the preceding event ε_{i-1}. Hence

$$d_i = f_i \left(c, \sum_{j=1}^{i-1} d_j \right) \equiv \psi_i(c) . \tag{17}$$

L_4^*: The total fatigue damage D is the sum of those damages per event ε_i, which exceed the threshold

$$d_0 = \psi_0(c_0) . \tag{18}$$

Hence,

$$D = \sum_{i=1}^{N} (d_i - d_0) = \sum_{i=1}^{N} [\psi_i(c) - \psi_0(c_0)] . \tag{19}$$

L_5^*: Due to the first application of the event ε_i, the specimen suffers an initial damage D_0, which in correspondence to (19) may be assumed as

$$D_0 = \sum_{i=1}^{N_0} [\psi_i(c) - \psi_0(c_0)] , \tag{20}$$

where N_0 is an appropriately chosen number.

Concerning equation (17) in L_3^*, the following remark may be useful: Substituting in (17) $d_0 = f_0(c)$, $d_1 = f_1(c,d_0) = f_1(c,f_0(c)) = \psi_1(c)$, $d_2 = f_2(c,d_0+d_1) = f_2(c,f_0(c)+\psi_1(c)) = \psi_2(c)$, etc., one realizes, that relationship (17) can be reduced as has been done before, to

$$d_i = \psi_i(c) . \tag{21}$$

In (21), the properly chosen functions ψ_i account for aging of the metal, expressed by the various and different functions f_i, as well as for the memory effect, represented in (17) by the dependence of d_i on the previously accumulated damage $D_{i-1} = \sum_{j=1}^{i=1} d_j$.

On the basis of axioms L_1^* to L_5^*, and using (21) instead of (17), the total damage done to the specimen can be calculated by means of

$$D+D_0 = 2 \sum_{i=1}^{N_0} [\psi_i(c)-\psi_0(c_0)] + \sum_{i=N_0+1}^{N} [\psi_i(c)-\psi_0(c_0)] . \tag{22}$$

The lifetime N follows from (22), if quantities N_0, D, D_0, c, and c_0 as well as functions ψ_0 and ψ_i, $i = 1,2,3,\ldots$, are known. It will of course be a certain difficulty to determine the functions ψ_0 and ψ_i experimentally.

If in (22) sums are being replaced by integrals, one obtains

$$D+D_0 = 2 \int_0^{T_0} [\psi(c(t))-const.]dt + \int_{T_0}^{T} [\psi(c(t))-const.]dt .$$

This expression can be changed further by setting $\psi(c(t))-const. = f(c(t))$. The result is

$$D+D_0 = const. + \int_0^T f(c(t))dt , \qquad D = const. + \int_0^T f(c(t))dt . \tag{23}$$

Equation (23) corresponds to the one in [5], which had been derived by V.V. Bolotin in a different manner. Such correspondence is a welcome support for the hypotheses presented here.

Reconsider (19) under the assumption that

$$\psi_i(c) = d(c) + \phi_i(c) . \tag{24}$$

In (24), the term $\phi_i(c)$ represents the change in damage which takes place at the i-th loading step. Now,

$$D = \sum_{i=1}^{N} [\psi_i(c)-\psi_0(c_0)] = \sum_{i=1}^{N} [d(c)+\phi_i(c)] - N\psi_0(c_0) ,$$

and

$$D = Nd(c) + \sum_{i=1}^{N} \phi_i(c)-N\psi_0(c_0) = Nd(c) \left[1 + \frac{\sum_{i=1}^{N} \phi_i(c)}{Nd(c)} \right] - N\psi_0(c_0) . \tag{25}$$

Let

$$\alpha(c,N) = 1 + \frac{\sum\limits_{i=1}^{N} \phi_i(c)}{Nd(c)} , \qquad (26)$$

be the function which accounts for aging and memory. Set for formal reasons in addition

$$\psi_0(c_0) = d_0(c_0)\alpha_0(c_0) , \qquad (27)$$

which can always be done with adequately chosen quantities d_0 and c_0. Then, (25) changes into

$$D = Nd(c)\alpha(c,N) - Nd_0(c_0)\alpha_0(c_0) . \qquad (28)$$

Accordingly, one has

$$D_0 = N_0 d(c)\alpha(c,N_0) - N_0 d_0(c_0)\alpha_0(c_0) . \qquad (29)$$

Consequently,

$$D+D_0 = Nd(c)\alpha(c,N) + N_0 d(c)\alpha(c,N_0) - (N+N_0)d_0(c_0)\alpha_0(c_0) .$$

Yet, a number $\tilde{N}$ can be found so that

$$Nd(c)\alpha(c,N) + N_0 d(c)\alpha(c,N_0) = (N+N_0)d(c)\alpha(c,\tilde{N}) .$$

Therefore, also

$$D+D_0 = (N+N_0)[d(c)\alpha(c,\tilde{N}) - d_0(c_0)\alpha_0(c_0)] . \qquad (30)$$

Moreover, set

$$d_0(c_0)\alpha_0(c_0) = C , \qquad D+D_0 = K .$$

Then, (30) assumes the form

$$[d(c)\alpha(c,\tilde{N})-C] = \frac{K}{N+N_0} . \qquad (31)$$

with

$$d(c)\alpha(c,\tilde{N})-C = F[d^*(c)] , \qquad N+N_0 = N^* , \qquad (32)$$

one has finally

$$F[d^*(c)] = \frac{K}{N^*} .\qquad\qquad(33)$$

This result may be compared with equation (10). One realizes, that the term $F[d^*(c)]$ in (33) is a logical generalization of the term $(\sigma^*)^{\beta}$. With (33), one has therefore again arrived at a somehow "deformed hyperbola", which is shown in Figure 3.

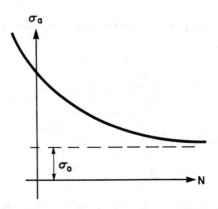

Figure 2 - Experimentally verified Wöhlercurve: $(\sigma_{a}-\sigma_0)^{\beta} = K(N+N_0)$

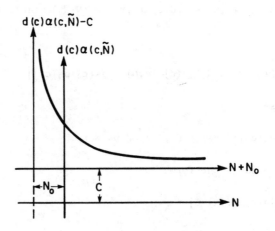

Figure 3 - Hypothetical Wöhlercurve according to equation (31), involving the effect of aging and memory displayed by the metal.

At the same time, one has provided with this comparison some justification for the empirical formula (10): One of the possible and most simple realizations of the term F[d*(c)] may be example be the term $(\sigma*)^\beta$, which can now be recognized as a means for expressing by a proper choice of the exponent β the aging and memory effect displayed by the metal under investigation.

The next considerations are devoted to loading with varying intensity. The simplest case of such loading is shown in Figure 4.

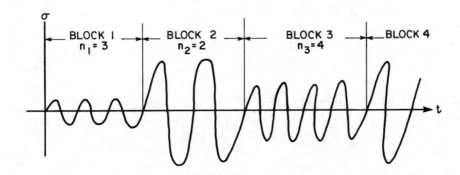

Figure 4 - Loading with Varying Intensity

Let it first be assumed, that one has loading blocks B_i, i = 1,2,3,..., consisting of n_i blockwise equally damaging events e_i with damage d_i per event. Let the block B_i have n_i events. Then, the damage per loading block B_i is

$$D_i = n_i d_i . \tag{34}$$

The total fatigue damage, disregarding for the sake of simplicity any initial damage, is

$$D = \sum_i D_i , \tag{35}$$

and the fatigue life is

$$N = \sum_i n_i . \tag{36}$$

The frequency of events e_i is given by

$$\beta_i = n_i/N , \tag{37}$$

from which follows

$$n_i = \beta_i N .$$ (38)

Fatigue damage can also be achieved by performing exclusively events e_k, $k \in \{i\}$, N_k being the fatigue life in this case. Hence,

$$D = N_k d_k , \qquad d_k = D/N_k .$$ (39)

Using (34) and (39) in (35) yields

$$D = \sum_i \frac{n_i}{N_i} D$$

or

$$1 = \sum_i \frac{n_i}{N_i} .$$ (40)

Substituting (38) into (40) results in

$$1 = \sum_i \frac{\beta_i}{N_i} N , \qquad 1 = N \sum_i \left[\frac{\beta_i}{N_i}\right] ,$$

from which follows

$$N = \sum_i \left[\frac{\beta_i}{N_i}\right]^{-1} .$$ (41)

This is Miner's rule, which has therefore been shown to be exact under the assumption that the loading can be described in terms of blocks B_i made up of physically equal, i.e., *equally damaging events*.

In practice, one may have difficulties operating with physically equal events. As already mentioned before, one may have to work rather with phenomenologically equal events, like load cycles. Then, the preceding considerations, which led to the classical form of Miner's rule, have to be modified as follows.

First, (34) have to be replaced by

$$D_i = \sum_{j=1}^{n_i} d_{i,j}$$ (42)

and, then, in accordance with (18), one may set

$$d_{i,j} = \psi_{i,j}(c) ,$$ (43)

which transforms (42) into

$$D_i = \sum_{j=1}^{n_i} \psi_{i,j}(c) \; . \tag{44}$$

Let (24) be generalized into

$$\psi_{i,j}(c) = d_i(c) + \phi_{i,j}(c) \; , \tag{45}$$

so that (44) can be changed into

$$D_i = n_i d_i(c) + \sum_{j=1}^{n_i} \phi_{i,j}(c) = n_i d_i(c) \left[1 + \frac{\sum_{j=1}^{n_i} \phi_{i,j}(c)}{n_i d_i(c)} \right] . \tag{46}$$

The function

$$\alpha_i(n_i,c) = 1 + \left[\sum_{j=1}^{n_i} \phi_{i,j}(c) \right] \, [n_i d_i(c)]^{-1} \; , \tag{47}$$

in (46) is supposed to account for aging and for the memory of preceding load blocks' influences.

With (47) in (46), one has simply

$$D_i = n_i d_i(c) \alpha_i(n_i,c) \tag{48}$$

for the contribution to damage by load block B_i.

Relations (35), (36), (37) and (38) remain unchanged. In correspondence with (48), relation (39) has to be replaced by

$$D = N_k d_k(c) \alpha_k(c,N_k) \; , \qquad d_k(c) = D/N_k \alpha_k(c,N_k) \; , \qquad k \in \{i\} \; . \tag{49}$$

Using the second one of the relationships in (49) for (48) yields

$$D_i = n_i \frac{D}{N_i} \frac{\alpha_i(n_i,c)}{\alpha_i(N_i,c)} \; . \tag{50}$$

Let the function

$$H(n_i,N_i,c) = \frac{\alpha_i(N_i,c)}{\alpha_i(n_i,c)} \; , \tag{51}$$

be introduced, so that (50) can be written as

$$D_i = n_i \frac{D}{N_i H(n_i,N_i,c)} \; . \tag{52}$$

Finally, set

$$N_i H(n_i,N_i,c) = G(n_i,N_i,c) \; , \tag{53}$$

which condenses (52) further into

$$D_i = n_i \frac{D}{G(n_i, N_i, c)} \, . \tag{54}$$

Now, let (54) be used in (35). Consequently, one is lead to

$$D = \sum_i \frac{n_i}{G(n_i, N_i, c)} D$$

and

$$1 = \sum_i \frac{n_i}{G(n_i, N_i, c)} \, . \tag{55}$$

Substituting (38) into (55) results in

$$1 = \sum_i \frac{\beta_i}{G(\beta_i, N, N_i, c)} N \tag{56}$$

from which follows

$$N = \Phi(\beta_i, N_i, c) \, . \tag{57}$$

This is a relationship, which replaces the simple Miner's rule (41). In order to work with (57), one had to proceed as follows. Equation (33), i.e.,

$$F(d^*(c)) = K/(N+N_0) \, , \tag{33}$$

may be transformed into

$$N = \Psi[d(c)] \, . \tag{58}$$

Every load block B_i is characterized by a distinct value $d_i(c)$. Using it in (58) yields

$$N_i = \Psi[d_i(c)] \, . \tag{59}$$

This result is substituted into (57) to yield

$$N = \Phi\{\beta_i, \Psi[d_i(c)]\}. \tag{60}$$

Since the quantities β_i and $d_i(c)$ are known through the definition of the loading process, equation (60) furnishes a value for the sought after lifetime N.

3. PRACTICAL APPLICATION OF DAMAGE ACCUMULATION THEORY

It may be rather difficult to base any practical application of the preceding damage accumulation theory on the generalized Miner formula (57). Function G involved in the preceding equation (56) had to be determined experimentally in a painstaking way. That appears to be a monumental task. However, experimental evidence seems to indicate that the situation is simpler than expected: the term $\alpha_i(n_i,c)$, appearing in (50), proves to be approximately a constant in a number of practical cases. This comes about by the various loading blocks B_i being relatively small so that the n_i are small numbers. With this fact in mind, one can change (51) into

$$H^* = \alpha_i^*(N_i,c)$$

in order to come up, instead of (53), with

$$N_i H^* = G^*(N_i,c) \ .$$

Consequently, one has in place of (54) the relationship

$$D_i = n_i \frac{D}{G^*(N_i,c)} \ . \tag{61}$$

Using (61) in (35), eliminating n_i by means of (38), and cancelling the common factor D, one obtains the equation

$$1 = \sum_i \frac{\beta_i}{G^*(N_i,c)} \ N \ . \tag{62}$$

From (62) follows easily

$$N = \left[\sum_i \frac{\beta_i}{G^*(N_i,c)} \right]^{-1} , \tag{63}$$

which is a more straightforward generalization than (57) of the elementary form (41) of Miner's rule.

Working with (63) in order to assess fatigue life N consists of the following procedure: from a generalized Wöhlerhyperbola (33), which may have been obtained experimentally, one derives as shown before relationship (59). From it, the various values N_i can be read off, which correspond to the values $d_i(c)$ as these are prescribed by the loading history. Using those values N_i in (63) yields for the fatigue life N the result

$$N = \left[\sum_i \frac{\beta_i}{G^*[\Psi(d_i(c))]} \right]^{-1} .$$ (64)

In order to make (63), (64), respectively, applicable in practice, one has to have a notion of the function G^*, while function Ψ had been derived earlier when transforming the known Wöhlercurve (33). Such notion of G^* can be obtained by choosing various classes of $\{\beta_i\}$ sets, simulating for these classes matching load histories with appropriate $d_i(c)$ values associated, and determining experimentally corresponding fatigue lifes N. Using these values of β_i, $d_i(c)$, and N in (64) may allow one to conclude on G^*. Yet, one is then, of course, facing an underdeterminate problem. Hence, one has to add certain assumptions on the possible structure of function G^*, which may however come up very naturally from insight into the physical nature of the fatigue process.

As will be shown subsequently by an example, it may be more convenient from the procedural point of view to follow a slightly different line of action. As a result of experiments, a Wöhlercurve is supposed to have been established. Consequently, one is provided with (59), i.e., $N_i = \Psi[d_i(c)]$. Now, apply operation G^* in order to obtain

$$G^*(N_i) = G^*\{\Psi[d_i(c)]\} .$$ (65)

Using the notation

$$\hat{N}_i = G^*(N_i) , \qquad G^*\{\Psi[d_i(c)]\} = \Psi^*[d_i(c)] ,$$ (66)

one has instead of (65)

$$\hat{N}_i = \Psi^*[d_i(c)] .$$ (67)

The graph of this expression can be interpreted as an additionally deformed Wöhlercurve. In view of (66), (67), one can change (64) into

$$N = \sum_i \left[\frac{\beta_i}{\hat{N}_i} \right]^{-1} ,$$ (68)

which is formally identical to (41), i.e., to the classical Miner rule. The difference between (41) and (68) is that (41) can be operated with the experimentally established Wöhlerhyperbola yielding the values N_i, while (68) has to be operated with the values $\hat{N}_i$, which are obtained from the modified Wöhlercurve (67).

Another, practically important point emerges due to the fact that
loading with varying intensity is frequently of a stochastic nature. Then,
equation (68) remains formally valid, but, the β_i have to be replaced by
the probabilities p_i of the occurrence of the loading blocks B_i in the
load history, and N as well as $\hat{N}_i$ have to be substituted by the respective
expectations E(N) and $E(\hat{N}_i)$ of these quantities. Hence, in the case of a
stochastic loading, one has to use the formula

$$E(N) = \left[\sum_i \frac{p_i}{E(\hat{N}_i)} \right]^{-1} . \tag{69}$$

The justification of (69) is obtained as follows. From (61),

$$E(D_i) E[G^*(N_i)] = E(n_i) E(D)$$

and

$$E(D_i) E(\hat{N}_i) = p_i E(D) E(N) ,$$

$$E(D_i) = \frac{p_i E(D) E(N)}{E(\hat{N}_i)} , \tag{70}$$

can be derived, using here and subsequently the assumptions that the
expectation of a product, is the product of the expectations of its
factors and that the expectation of a sum is the sum of the expectations
of its terms.

From (35), one has

$$E(D) = \sum_i E(D_i) . \tag{71}$$

Using (70) in (71) yields

$$E(D) = \sum_i \frac{p_i E(D) E(N)}{E(\hat{N}_i)} ,$$

which can easily be changed into

$$E(N) = \left[\sum_i \frac{p_i}{E(\hat{N}_i)} \right]^{-1} .$$

This result is identical to (69).

4.   AN EXAMPLE

Based on experiments designed and carried out by M. El Menoufy [6], the fatigue life evaluation for stochastic loading using the concept of modified Wöhlercurves has been presented in [7]. In [8], such considerations were repeated, stressing in addition the reasons for the emergence of a modified Wöhlercurve (67), and giving a notion of its appearance.

The load program consisted of four-strain-level cycles as shown in Figure 5. Such program involves 16 blocks of load cycles with the associated probabilities p_i, i = 1,2,...,16, as shown in [7]. As a damage parameter, the Smith-parameter

$$c = (\sigma_{max} \frac{\Delta \varepsilon}{2} E)^{1/2} , \tag{71}$$

advocated in [4], was chosen. For a prescribed set of probabilities p_i, load programs were simulated by means of the Monte Carlo technique yielding a set of Wöhlercurves. Since cycles had been chosen as damage events, which are in general not equally damaging, one had obtained Wöhlercurves of the type (31), i.e.,

$$[d(c)\alpha(c,\tilde{N})-C] = \frac{K}{N+N_0} ,$$

$$F[d^*(c)] = \frac{K}{N^*} , \tag{33}$$

$$d(c) = F(N) \tag{72}$$

respectively, see Figure 6.

The next step consisted in applying operator G* to N in order to produce the modified curve

$$d(c) = F[G^*(N)] = F^*(N) , \tag{73}$$

which is shown in Figure 7. This curve was actually not derived analytically but empirically, so that operator G* was never established explicitly. The operation attributed to G* consisted in replacing, from a certain value N_c on, the curve (72) by its tangent at N_c.

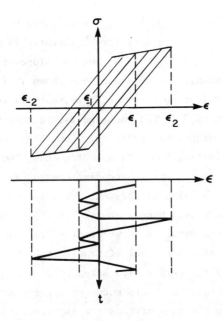

Figure 5 - *Stochastic Load Program with Four Strain Levels*

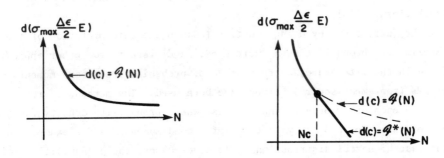

Figure 6 - *Experimentally Obtained   Figure 7 - Modified Wöhlercurve*
Wöhlercurve

Justification of this operation is the fact that the sufficiently
often occurrence of cycles with sufficiently large *compressive* peaks
would cause cycles with small peaks to be more damaging than they would
be, if they were applied exclusively and without any memory of preceding
large compressive peaks. Hence, these small peak cycles would have
smaller lives than given by the Wöhlercurve in Figure 6. Such decrease

of fatigue life for small peak cycles would then be taken care of by the
replacement of the lower part of the Wöhlercurve' in Figure 6 by its tangent
at N_c as shown in Figure 7. How N_c and the slope of the tangent can be
found, using the condition (69), has been shown in [8].

A lesson to be learned from the experiments is that for some metals,
like the one being used in the experiments, the memory effect is limited
to selected, preceding load cycles only, but the effect of the selected
cycles is a very lasting one. As a consequence, the operator G* account-
ing for aging due to memory of preceding damage, is a rather simple one,
which is in addition fairly invariant over a wide range of load histories.
All these facts have also been discussed in [8], where the reason for
this response of the metal to large compressive peaks has been explained
as being a result of the removal of crack-closure. Such unlocking of
crack-closure by large compression makes it in fact possible for subsequent cycles with small peaks to make greater contributions to damage
than they would in the case of constant-amplitude loading.

The phenomenon of crack-closure has been described for example by
J.C. Newman, Jr. [9], and T. Topper and P. Au have reported on the long
lasting effect which large compressive cycles have on the removal of
crack-closure [10].

The fact, that, by using equation (69) in the interpretation of the
experiments, probabilities p_i were used, is different from other approaches
found in the literature, where instead of probabilities second moments
or RMS (root-mean-square) values have been used. The methods correspond to each other only in those cases, where probability density functions
are expressible accurately in terms of second moments as for example in
the case of normal distributions. In other cases, the probabilities for
the load blocks of the loading with varying intensity have to be determined in preliminary calculations as for example described in [8].

Results of four-strain level fatigue tests carried out by El Menoufy
on the basis of equation (69) are reported in Table 1. Probabilities
π_i, i = -2,-1,1,2, in this table are the probabilities of occurrence of
the peak strains ε_{-2}, ε_{-1}, ε_1, ε_2 of the load cycles. It has been shown
in [7] and [8] how these π_i - values can be used to calculate the cycle
probabilities p_i, i = 1,2,...,16. The results shown in Table 1 indicate
clearly the good correspondence of calculated values E[N] with experimental
realizations of N.

Table 1 - *Fatigue Life Expectations for Random Loading with with Four Strain Peaks*

Test #	Probability Distribution				N experimental	E(N) predicted
	π_{-2}	π_{-1}	π_1	π_2		
Strain Levels $\varepsilon_{-2}=-0.3\%, \varepsilon_{-1}=-0.1\%$ $\varepsilon_1 =+0.1\%, \varepsilon_2 =+0.3\%$						
1	0.25	0.20	0.30	0.25	157,476	112,353
2	0.10	0.30	0.20	0.40	131,500	106,798
3	0.40	0.20	0.30	0.10	148,989	163,492
Strain Levels $\varepsilon_{-2}=-0.4\%, \varepsilon_{-1}=-0.1\%$ $\varepsilon_1 =+0.1\%, \varepsilon_2 =+0.4\%$						
4	0.15	0.30	0.35	0.20	49,486	55,834
5	0.10	0.30	0.20	0.40	50,648	42,550
6	0.40	0.20	0.30	0.10	73,209	73,795
7	0.40	0.20	0.30	0.10	83,728	80,315
8	0.10	0.30	0.20	0.40	59,839	46,365
Strain Levels $\varepsilon_{-2}=-0.5\%, \varepsilon_{-1}=-0.1\%$ $\varepsilon_1 =+0.1\%, \varepsilon_2 = +0.5\%$						
9	0.25	0.20	0.30	0.25	21,803	21,978
10	0.15	0.35	0.35	0.15	42,830	39,580
11	0.25	0.25	0.25	0.25	25,960	23,630

ACKNOWLEDGEMENT

The support of this research by NSERC through Grant No. A7297 is gratefully acknowledged.

REFERENCES

1. WEIBULL, W., *Fatigue Testing and Analysis of Results*, 1961, p. 174.
2. MASON, S.S., *Tech. Note 239*, Nat. Advis. Comm. Aero., 1954.
3. COFFIN, L.F., *Trans. ASME*, Vol. 76, 1954, p. 923.
4. SMITH, K.N., WATSON, P. and TOPPER, T.H., "A Stress-Strain Function for the Fatigue of Metals", *Journal of Materials*, JMLSA5, 1970, pp. 767-778.
5. BOLOTIN, V.V., "Wahrscheinlichkeitsmethoden zur Berechnung von Konstruktionen", *VEB Verlag für Bauwesen*, Berlin, 1981, pp. 487-491.
6. EL MENOUFY, M., "Fatigue Life Evaluation for a Metal Subjected to a Certain Class of Random Loading Using the Modified Life-Law Concept", *M.A. Sc. Thesis*, University of Waterloo, 1982.
7. EL MENOUFY, M., LEIPHOLZ, H.H.E. and TOPPER, T.H., "Fatigue Life Evaluation, Stochastic Loading, and Modified Life Curves", *Shock and Vibration Bulletin 52*, Part 4, May 1983, pp. 11-19, Naval Research Laboratory, Washington, D.C.
8. LEIPHOLZ, H.H.E., TOPPER, T.H. and EL MENOUFY, M., "Lifetime Prediction for Metallic Components Subjected to Stochastic Loading", *Computers and Structures*, Vol. 16, 1983, p0. 499-507.

9. NEWMAN, J.C., Jr., "A Crack-Closure Model for Predicting Fatigue Crack Growth under Aircraft Spectrum Loading", *Methods and Models for Predicting Fatigue Crack Growth under Random Loading*, edited by J.B. Chang and C.M. Hudson, ASTM Publ. PCN04-748000-30, 1981, pp. 53-84.
10. TOPPER, T.H. and AU, P., "Cyclic Strain Approach to Fatigue in Metals", *AGARD-Lectures Series-118*, NTIS, pp. 11-1-11-25, Springfield, Virginia, 1981.

THE STATE OF AFFAIRS NEAR THE CRACK TIP

G.C. SIH
Institute of Fracture and Solid Mechanics
Lehigh University
Bethlehem, Pennsylvania, U.S.A.

1. INTRODUCTION

Pedagogically speaking, a crack tip is the end point of two inter-
secting lines. In reality, the tip region is irregular and highly influenced by material inhomogeneity and imperfection at the microscopic level.[1]
Its contour may follow the pores within a single grain or the broken segments of many grains. In the continuum model, however, the crack tip is
assumed to follow a smooth contour so as to reduce the mathematics to
manageable proportions. The influence of material microstructure cannot
be decided by geometric consideration alone. Attention should be focused
on the rate of local energy transfer that is governed by specimen size
and loading rate. The trade-off between these two effects is pertinent to
analyzing damage of the crack tip region.

A crack tends to disturb the local homogeneity in a system through its
tip region which attains a relatively high intensity of energy depending
on the nature of the loading. This influence can often be reflected globally through measurements of load and displacement as the material is
being damaged. A complete knowledge of the damage process, therefore,
necessitates the correlation and definition of local and global failure,
and the specification of size scale at which damage is being assessed both
qualitatively and quantitatively. Unless the relations between many of
the microscopic and macroscopic material damage models are better understood [1], little or no progress can be made to enhance prediction of
material and/or structure behaviour.

The degree of inhomogeneity that needs to be reflected through the
crack tip region depends on the local and global failure modes. In the

---

[1] Imperfections at the atomic level should be referred to as vacancies, dislocations, etc. In this communication, *crack* refers to material discontinuities at the microscopic and macroscopic level.

simplest case of brittle fracture, where local and global failure are
assumed to coincide, it suffices to consider complete homogeneity of the
crack tip region regardless of the microstructure. Either the breaking
of a microelement or macroelement ahead of the crack could lead to global
instability. Although the specific fracture surface energy or energy
required to break a microelement will differ from that of a macroelement,
both models should yield the same critical load at fracture. The correspondence between the micro- and macro-cracking model involves only a scale
shifting factor [2]. When yielding[2] occurs near a macrocrack tip, the
stress state becomes nonhomogeneous[3] in that the local moduli acquire a
nonuniform distribution. Global instability no longer occurs immediately
but is delayed by slow crack growth after initiation [3]. The damage process then becomes load history and specimen size dependent for a given
material. The influence of microstructure becomes more and more important
as the rate of energy input to the crack tip region is slowed down such
that the nonhomogeneous character of the grains begins to influence the
load transfer characteristics. Creep and/or fatigue failure are cases
in point. For the same microstructure, material can exhibit different
failure behaviour ranging from brittle fracture to plastic collapse or
excessive elongation. It is the rate of energy dissipated through a unit
volume of material that determines a particular failure mode.

The phenomenon of crack bifurcation is another example of how crack
tip material inhomogeneity [4] can affect failure mode. A crack becomes
increasingly more unstable as its speed of propagation is increased. It
can then be easily deviated from its original path of travel when encoun-
tering local inhomogeneity created by a void or defect. A second branch
is frequently generated from the point where the original crack has turned
because the first branch cannot dissipate sufficiently when it propagates
towards the direction of applied tension. This event occurs so quickly
that it gives the impression of crack bifurcation which is a direct consequence of material inhomogeneity.

_____

[2]
Macroyielding reflects material damage at the microscopic level and the
effects are smeared over a region in the continuum analysis referred to
as the plastic enclaves.
[3]
Nonuniform yielding around the crack tip invalidates the concept of elastic or plastic stress intensity factor since the stress field cannot be
uniquely characterized by a common field strength or stress singularity.
The assumption of uniform yielding around the crack is nonphysical and is no
consistent with experimental observation.

Crack tip detail can affect global material behaviour. This depends
on the combined effects of loading, specimen size and material type. The
local and global stationary values of the strain energy density function
dW/dV can be used to analyze the instability of a damaged system. Such a
general concept can be extended to account for inhomogeneity due to
material microstructure, a topic of challenge for future research.

2. MODELLING OF CRACK TIP REGION

The foundation of mechanics of deformable bodies relies on obtaining
material properties of the constituents in a continuum system from smaller
specimens that are tested under simple loading conditions. This procedure
is not always straightforward and frequently misapplied when energy dissipation due to material damage is involved. The range of size and time
scale within which damage could be assessed is presently limited to a few
orders of magnitude mainly because of the lack of *consistency* in treating
damage at the atomic, microscopic and macroscopic level. Events observed
at the different scale level are being reported mostly by empirical means
and not related systematically on a sound theoretical basis. In what
follows, discussions will be centred on modelling of the crack tip region
and the physics of material damage.

2.1 Core Region

The intensity of energy transferred to the crack tip depends largely
on the loading rate and the physical dimensions of the crack in relation
to the microstructure. In order to specify the size scale at which damage
is to be evaluated, it is necessary to consider a core region of radius
r_0 around the crack tip as a measure of the resolution of analysis. Suppose
that the crack tip radius of curvature ρ is small in comparison with the
mean linear dimension d of the grains in a polycrystal, i.e., $\rho \ll d$. The
crack tip is then likely to be situated in a single grain as illustrated
in Figure 1(a). To be analyzed is a submicroelement having the local properties of a grain. The details within the core region are assumed to
have negligible influence on the global response of the system. When $\rho \simeq d$
as indicated in Figure 1(b), the core region radius r_0 would have to be
increased accordingly and a microelement consisting of the average properties of the grains should be analyzed. The assumption of material

homogeneity prevails as ρ becomes many times greater than d, Figure 1(c). The dimension r_0 is then comparable with the macroelement size.

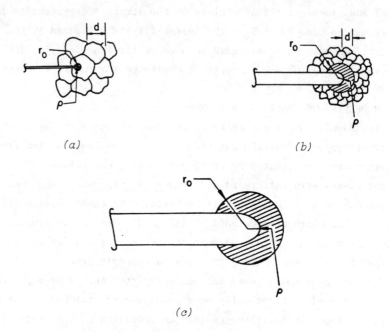

(a)   *(b)*

(c)

Figure 1 - *Size Crack Tip Compared with Material Microstructure*
  (a) crack tip in a grain: $\rho \ll d$
  (b) crack tip among grains: $\rho \simeq d$
  (c) crack tip in homogeneous material: $\rho \gg d$

The size of the core region r_0 around a notch or crack tip has been studied extensively in [5,6]. An estimate of r_0 in relation to the notch length was made by matching the crack trajectories emanating from a notch front with those from a sharp crack tip. The size r_0 corresponds to the location at which the two sets of trajectories coincide. It varies with load direction and type.

2.2 *Micro and Macro Damage*

Realistic modelling of material failure near the crack tip region requires a knowledge of damage at both the microscopic and macroscopic level. Consider the classical example of the apparent brittle fracture of a tensile crack in a low carbon steel as reported in [7]. X-ray detection, however, showed a thin layer of highly distorted material along the

fracture surface. This seemingly contradicting observation[4] can be clari
fied by referring to the schematic representation of the situation in
Figure 2. According to continuum mechanics analysis, the macroelements
along the path of prospective macrocracking experience macrodilatation and
microdistortion. This explains for the creation of slanted microcracks
prior to macrocracking along planes normal to the applied tension. Similarly, the macroelements off to the sides are subjected to macrodistortion
and microdilatation. The former is responsible for macroyielding and the
latter for microcracks in the plastic enclaves.

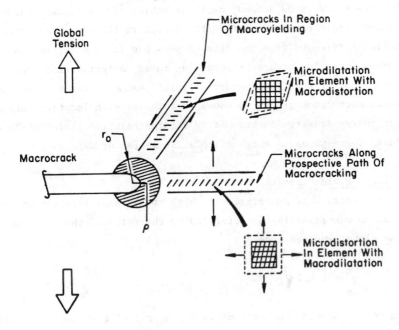

Figure 2 - Schematic of Macro- and Micro-Damage in the Region
Ahead of Crack Tip

In view of the above remarks, it is mandatory to consider the distortional and dilatational component of the energy stored in the macroelements ahead of the crack. They both play a role in the material damage
process [8]. The dilatational component dominates along the path of

---

[4] A quantity, γ_p, referred to as plastic work, was introduced in [7] to
account for the plastic deformation in a thin layer of material that failed
apparently in a brittle fashion. This explanation has been shown to be
inadequate [2] because of the failure to distinguish specific surface
energy for the creation of micro- and macrocrack surface.

macrocracking while the distortional component dominates at the sites of macroyielding. Referring to Figure 2, macrocracking is initiated by micro-distortion and macroyielding by microdilatation. In this respect, the von Mises yield condition is incomplete as it accounts only for the distortional component of the strain energy density function and cannot explain the creation of microcracks in the plastic zones.

On physical grounds, the strain energy density criterion [8,9] provides a more complete description of material damage in the crack tip region. Both the distortional and dilatational effects are included and their con-tributions are weighed automatically by taking the stationary values of the total strain energy density with respect to the appropriate space variables referenced from the site of possible failure initiation. The criterion is valid for cracks spreading in any material that behaves in a reversible or irreversible manner. Typical examples can be found in [10-12] in which macrocrack growth is shown to coincide with locations where the strain energy density function dW/dV attains relative minima while macro-yielding corresponds to sites of relative maxima of dW/dV.

2.3 Local Field Strength

The mathematical description of high stress magnification at the crack tip can be conveniently characterized by the ratio of the local stress σ_m to the global applied stress σ_g:

$$\frac{\sigma_m}{\sigma_g} = 1 + 2 \sqrt{\frac{a}{\rho}} , \tag{1}$$

which corresponds to the case of an elliptical slit of length 2a with ρ being the radius of curvature at the end of the major axis. If the slit is very narrow, then $a \gg \rho$ and equation (1) may be written as[5]

$$\sigma_g \sqrt{a} = \frac{1}{2} \sigma_m \sqrt{\rho} . \tag{2}$$

The quantity $\sigma_g \sqrt{a}$ is recognized as the stress intensity factor k_1 or the strength of the singular stress field. The field strength is said to be homogeneous in that all the local stress components σ_{ij} (i,j = x,y) possess the same $1/\sqrt{r}$ singularity:

[5]
It is understood that although k_1 is the coefficient of the $1/\sqrt{r}$ singular stress field, the stress intensity factor concept applied to a finite crack with a radius of curvature $\rho = 4a(\sigma_g/\sigma_m)^2$.

$$\sigma_{ij} = \frac{k_1}{\sqrt{2r}} \, f_{ij}(\theta) \, , \tag{3}$$

in which r and θ are the polar coordinates defined in Figure 3(a).

When yielding occurs, the singular character of the stress field becomes nonhomogeneous. Although equation (3) remains valid in the elastic region, the stress components σ^*_{ij} in the plastic zone will acquire a singularity of the type $1/r^\lambda$ such that λ is different from 1/2. Hence, there no longer prevails an unique coefficient that can characterize the strength of the elastic-plastic stress field, Figure 3(b). The selection of any stress or strain parameter to be used as a failure criterion would therefore become arbitrary.

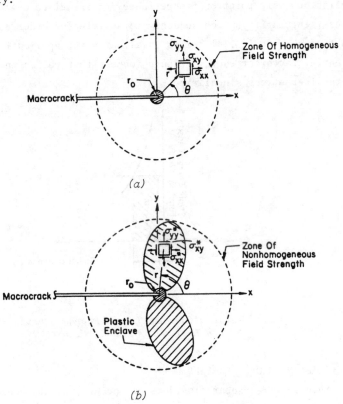

(a)

(b)

Figure 3 - Elastic and Elastic-Plastic Stress Field
Near Crack Tip

(a) elastic deformation
(b) elastic-plastic deformation

It is of fundamental interest to know that the strain energy density field always possess the 1/r singularity even though the stress field characteristics may be nonhomogeneous. This character holds for all linear nondissipative and nonlinear dissipative materials within the framework of continuum mechanics. Referring to Figure 4, dW/dV can, in general, be written as

$$\frac{dW}{dV} = \frac{S}{r} \; . \tag{4}$$

The strain energy density factor S is a finite quantity defined by the area under the curve at a distance r. The character of equation (4) also applies to every point on the periphery of a surface crack whose $1/\sqrt{r}$ stress singularity switch suddenly to a different order at the corner of the free surface [13,14]. This invalidates the application of the stress intensity factor or energy release concept that relies on the existence of the $1/\sqrt{r}$ stress singularity.

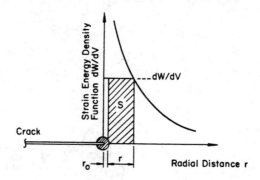

*Figure 4 - Variations of Strain Energy Density Function
with Distance Ahead of Crack*

3. LOCAL AND GLOBAL FAILURE

Since material separation does not occur instantaneously, local failure should, in general, be distinguished from global failure when classifying the different types of material damage modes. Common termino-logies such as ductile fracture, fatigue crack propagation and creep failur all describe subcritical crack growth and/or irreversible deformation prior to complete termination of structural integrity. Analytical modellin of these physical damage processes requires a rational approach that can consistently describe damage modes at the different scale level.

3.1 Brittle Behaviour

Brittle material has been customarily referred to the linear elastic behaviour of an uniaxial tension specimen broken suddenly with little permanent deformation. The same material, however, can behave in a very ductile fashion simply by changing loading rate, specimen size or temperature. Similarly, a material that exhibits ductile behaviour in uniaxial tests can also be made to behave in a brittle manner if the test conditions are altered. In order to avoid ambiguity, the local and global view of failure will be invoked.

Brittle behaviour can be best described in terms of the coincidence of local and global failure. In the case of the Griffith crack in Figure 5, a crack of length 2a is embedded in a field of uniform tensile stress σ_g. The condition of incipient brittle fracture is given by

$$\sigma_g = \sqrt{\frac{2E}{\pi a}} \sqrt{\gamma_g} \, , \tag{5}$$

where γ_g is the energy required to create a unit of macrocrack surface, say $\Delta L \cdot 1$, Figure 5. All the energy is assumed to be released at the instant of global instability that corresponds to the breaking of a macroelement. By means of equation (2), equation (5) may be expressed in terms of the local stress σ_m as

$$\sigma_m = \sqrt{\frac{8E}{\pi \rho}} \sqrt{\gamma_g} \, . \tag{6}$$

Alternatively, the same global failure stress σ_g can be predicted from the brittle fracture of a microelement. The energy required to create a unit of microcrack surface, say $\Delta \ell \cdot 1$ (Figure 5), is γ_p. The local microstress σ_n takes a form similar to σ_m in equation (6):

$$\sigma_n = \sqrt{\frac{8e}{\pi d}} \sqrt{\gamma_p} \, , \tag{7}$$

except that e is the microscopic modulus of elasticity and d the mean length dimension characterizing the microstructure. The macroscopic parameters σ_m and γ_g cannot, in general, be equated to the microscopic parameters

σ_n and γ_p. A relation between equations (6) and (7) can be derived by assuming that the same critical *force* acting on the micro- or macroelement causes incipient fracture:

$$f_m = f_n \quad \text{or} \quad \sigma_m \cdot \Delta L \cdot 1 = \sigma_n \cdot \Delta \ell \cdot 1 . \tag{8}$$

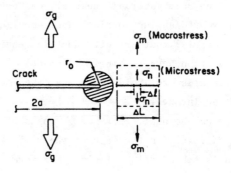

Figure 5 - Macro- and Micro-Model of Brittle Fracture

The condition of critical *total energy* at fracture leads to

$$2\gamma_g \cdot \Delta L \cdot 1 = 2\gamma_p \cdot \Delta \ell \cdot 1 , \tag{9}$$

which is equivalent to equation (8) because

$$\frac{\Delta L}{\Delta \ell} = \frac{\sigma_n}{\sigma_m} = \frac{\gamma_p}{\gamma_g} \simeq 10^3 \text{ to } 10^4 \text{ cm } . \tag{10}$$

Force and total energy are quantities that can be used to bridge the gap between macroscopic and microscopic material damage models. The quantity $\Delta L/\Delta \ell$ represents the ratio of macrolength to microlength and can be taken approximately as 10^3 to 10^4 cm. It is now clear that the specific micro-surface energy γ_p is roughly 10^3 to 10^4 times greater than the specific macrosurface energy γ_g in the Griffith's equation. This explains the measurement of γ_p in [7] and removes the ambiguity of plastic work done in a thin layer of material failed by brittle fracture.

Making use of equations (6), (7) and (10), the Griffith's critical stress σ_g in equation (5) may also be expressed in terms of γ_p since

$$\frac{\gamma_p}{\gamma_g} = \left(\frac{e}{E}\right)\left(\frac{\rho}{d}\right). \tag{11}$$

This gives

$$\sigma_g = \sqrt{\frac{2E}{\pi a}} \cdot \sqrt{\left(\frac{d}{\rho}\right)\left(\frac{E}{e}\right)} \sqrt{\gamma_p}. \tag{12}$$

Note that $\sqrt{Ed/e\rho}$ is a scale shifting factor and was not accounted for[6] in [7].

It is now well-known that brittle fracture phenomenon can be associated with several other parameters such as the critical energy release rate G_{1c}, stress intensity factor K_{1c} and strain energy density factor S_c, i.e.,

$$2\gamma_g = G_{1c} = \frac{(1-\nu^2)K_{1c}^2}{E} = \frac{2\pi(1-\nu)}{1-2\nu} S_c. \tag{13}$$

In particular, the relation

$$\left(\frac{dW}{dV}\right)_c = \frac{S_c}{r_c}, \tag{14}$$

allows the determination of S_c from area under the true stress and strain curve $(dW/dV)_c$ [15] and r_c is the critical ligament that triggers brittle fracture.

3.2 Ductile Behaviour

Ductile behaviour is characterized by substantial permanent deformation of the solid prior to fracture. The energy dissipation rate in the crack tip region is conducive to creating microcracks in the grains off to the side of the main crack as illustrated in Figures 6 and 7. This reduces the crack driving force and results in subcritical crack growth that can lead to sudden fracture or plastic collapse of the solid depending on the rate at which energy is transferred to the unbroken ligament. In either case, local failure precedes global instability. They coincide only at the

---
[6] It is not justified to add γ_p onto γ_g [7] and then drop γ_g on the basis that $\gamma_p \gg \gamma_g$, i.e.,

$$\sigma_g = \sqrt{\frac{2E}{\pi a}} \sqrt{\gamma_g + \gamma_p} \simeq \sqrt{\frac{2E}{\pi a}} \sqrt{\gamma_p}.$$

The scale shifting factor in equation (12) is not unity.

instant when the system ceases to support additional load. Figure 6(a)
shows macrocrack growth along a path of elastic material with yielding
off to the sides. The final fracture can be brittle if sufficient energy
is stored and released in the last ligament. On the other hand, the
material along the crack path can be yielded and the material in the last
ligament becomes highly distorted, Figure 7(a). The specimen may then
collapse before fracturing. These failure modes are not uncommon in
practice.

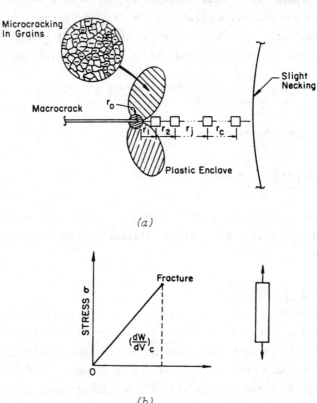

(a)

(b)

*Figure 6 - Crack Growth in Elastic Portion of Elastic-Plastic
          Stress Field*

      (a) crack in elastic ligament
      (b) strain ε linear elastic response

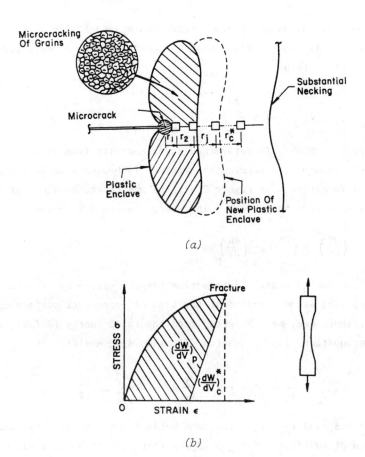

(a)

(b)

Figure 7 - Crack Growth with Yielding Along the Path
  (a) crack growth in yielded material
  (b)

Even though the macrocrack in Figure 6(a) spreads in the elastic region, the local stress redistribution must be determined from an elastic-plastic analysis so that the available energy to drive the crack may be found. This rate process is assumed to be governed by the relation [8]:

$$\left(\frac{dW}{dV}\right)_c = \frac{S_1}{r_1} = \frac{S_2}{r_2} = \cdots = \frac{S_j}{r_j} = \cdots = \frac{S_c}{r_c} \text{ or } \frac{S_0}{r_0} = \text{const.} \qquad (15)$$

The condition $S_c/r_c = \text{const.}$ corresponds to

$$S_1 < S_2 < \cdots < S_j < \cdots < S_c \,, \qquad (16)$$
$$r_1 < r_2 < \cdots < r_j < \cdots < r_c \,,$$

where the material damage process increases monotonically up to global instability. If the material damage process comes to arrest, i.e., S_0/r_0 = const., then

$$S_1 > S_2 > \cdots > S_j > \cdots > S_0 \,,$$

$$r_1 > r_2 > \cdots > r_j > \cdots > r_0 \,. \tag{17}$$

The quantity $(dW/dV)_c$ is determined experimentally from an uniaxial tensile test, Figure 6(b), where the specimen is broken with little distortion or deformation. In Figure 7(a), the macroelements ahead of the crack are yielded before they break. The energy released is hence given by

$$\left(\frac{dW}{dV}\right)_c^* = \left(\frac{dW}{dV}\right)_c - \left(\frac{dW}{dV}\right)_p \,. \tag{18}$$

Energy loss due to plastic deformation $(dW/dV)_p$ given by the shaded area in Figure 7(b) is not present at the time of macrocrack surface creation. It must, therefore, be subtracted from the total energy $(dW/dV)_c$ as indicated in equation (18). Equation (15) should be modified as

$$\left(\frac{dW}{dV}\right)_c^* = \frac{S_1}{r_1} = \frac{S_2}{r_2} = \cdots = \frac{S_j}{r_j} = \cdots = \frac{S_c^*}{r_c^*} \text{ or } \frac{S_0^*}{r_0^*} = \text{const.,} \tag{19}$$

where $S_c^* < S_c$ and $r_c^* < r_c$. The same holds for S_0^* and r_0^*. The intermediate situation of $(dW/dV)_c^* = S_c/r_c$ or S_0/r_0 that lies in between equations (15) and (19) can also arise. This corresponds to crack growing in the yielded material but results in brittle fracture characterized by S_c.

The aforementioned concept has been applied to solve a number of ductile fracture problems [10-12] that verified the assumption dS/da = con. Figures 8(a) to 8(c) give typical results in the S versus a plots. Varied in the stress and failure analysis are specimen size, loading rate and material type in terms of yield strength and fracture toughness. For a fixed material and loading rate, the smaller specimens are seen to sustain a more subcritical damage while the larger specimens have a tendency to fail catastrophically in a brittle fracture mode. Failure by different degrees of yielding and fracture are determined by parallel lines in the S versus a plot, Figure 8(a). The effects of loading rate can be assessed by a counterclockwise rotation of the S-a lines that intersect at a common point in Figure 8(b). For a constant S level, lower loading rates can

enhance slow and stable crack growth while high loading rates promote
sudden cracking. The range of specimen sizes for covering the same
failure modes in Figure 8(a) can be extended as the loading rate is
reduced. By increasing the area under the true stress and strain curve
or $(dW/dV)_c$, the S-a lines again rotate counterclockwise. This means
that materials with higher $(dW/dV)_c$ values when tested under the same
conditions tend to fail with less crack growth as compared with inelastic
deformation.

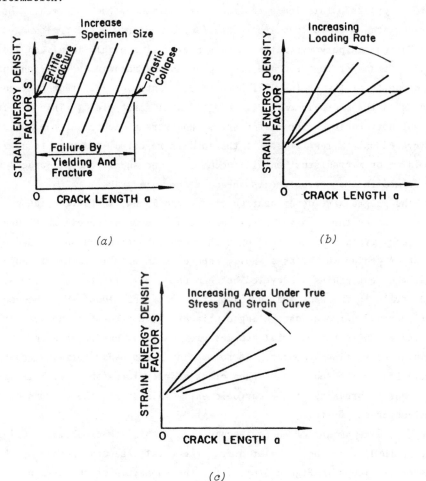

(a)

(b)

(c)

*Figure 8 - Schematic of Resistance Curves for Changes in Specimen
Size, Loading Rate and Material Toughness*

    (a) change in specimen size
    (b) change in loading rate
    (c) change in material toughness

3.3 Fatigue Crack Growth

As the applied stress amplitude is lowered below that of yield strengt and repeated periodically, the material ahead of the crack will be deterio ated in a progressive manner. This process is known as *fatigue*. In contrast to ductile fracture, damage of the microstructure does not occur on the first cycle of loading, rather localized damage in the form of submicroscopic cracks are developed after many cycles of repeated loading. Eventually, the damage becomes visible to the naked eye and grows in the form of cracks. This leads to final destruction of the specimen. The various stages of fatigue are *initiation, crack growth* and *fracture*.

Fatigue in polycrystalline metals begins with highly localized yieldin As load transfer to the crack tip region is now relatively slow, the non-homogeneity of the grains will have time to interact such that the weak grains in the macroelement, Figure 9(a), will first yield. This effect is not noticeable on the static stress and strain diagram if the material is cycled only a few times. If the load is repeated, however, localized yielding or microdistortion will accumulate and submicroscopic cracks will be formed in the macroelements ahead of the main crack.

The progressive deterioration of a macroelement in fatigue can be visualized by the uniaxial stress and strain reversal scheme in Figure 9(b The weak crystals in the zone of high microdistortion are surrounded by stronger grains which, as a whole, behave in an elastic fashion. Suppose that the weak grains are cycled between the strain limits $\pm\varepsilon_0$. The stress and strain in the weak grains follow the hysteresis loops which become narrower and narrower as cycling continues. The stress in the weak grains increases while the material becomes progressively harder as a result of cumulative strain-hardening. At some point, slip bands form and grow into striations. The weak grains then crack at the microscopic level leading to eventual breaking of the macroelement. This starts the process of fatigue crack growth.

The amount of energy accumulated within a macroelement during cyclic loading can be calculated from the total area transversed by the hysteresis loops in Figure 9(b), i.e., the summation of the strain energy density function:

$$\sum_{j=1}^{n} \left(\frac{dW}{dV}\right)_j = A = \text{const.}, \tag{20}$$

where n is the number of stress and/or strain reversal cycles to break
the macroelement. An average hysteresis energy density $(\Delta W/\Delta V)_{ave}$ may
be defined in terms of the interval number of cycles Δn as

$$\left(\frac{\Delta W}{\Delta V}\right)_{ave} \Delta n = A , \tag{21}$$

where A is obtainable from an uniaxial fatigue test of a smooth specimen.
Knowing that $\Delta W/\Delta V$ is equal to $\Delta S/\Delta a$, a fatigue growth relation is obtained:

$$\frac{\Delta a}{\Delta N} \left(\frac{\Delta N}{\Delta n}\right) = C(\Delta S) . \tag{22}$$

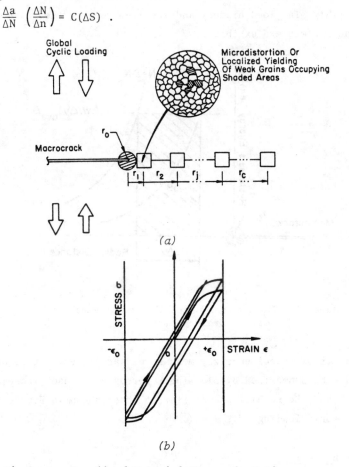

(a)

(b)

Figure 9 - Localized Material Damage in Fatigue
(a) localized yielding of weak grains
(b) hysteresis loops experienced by
*   macroelements*

The ratio $\Delta N/\Delta n$ is unity if the interval of applied load cycle ΔN and cyclic response Δn experiences by the local macroelements possess a one-to-one correspondence. The constant C is the y-intercept on a log $\Delta a/\Delta N$ versus log ΔS plot once the fatigue crack growth for a given material is known. An independent verification of the relation C = 1/A can then be made with A measured independently. For each increment of crack growth, ΔS must be computed and represents the amount of energy dissipated as the specimen is cycled from N to N+ΔN, Figure 10. This procedure has been applied to investigate fatigue crack growth in specimens with a centre crack [17] and edge crack [18]. The load history and specimen geometry dependence of fatigue crack growth are exhibited.

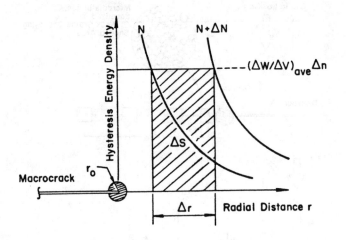

Figure 10 - *Energy Dissipated for an Increment of*
Fatigue Crack Growth

The effects of specimen size, mean stress and material type on fatigue life can also be summarized by plotting S versus a for each increment of crack growth and the results are similar to those shown in Figure 8 for the case of monotonic loading.

3.4 Creep Behaviour

Creep is more difficult to analyze than other material failure behaviour because it involves prediction of long-time behaviour from short-time test results. The mechanism of energy dissipation becomes more sensitive to the microstructure as loading rate is slowed down appreciably. The polycrystalline metals begin to lose their ability to strain-harden. They may fracture following a nearly uniform elongation without forming a neck, i.e., with very little plastic deformation. This behaviour can be identified with thermal softening where inhomogeneities at the grain boundaries become important. The rate of energy transfer to the crack tip region enhances movement of whole grains relative to each other. Permanent reorientation of the grains introduces microcracks at the boundaries because of grain shape irregularities, Figure 11(a). The coalescence of these microcracks results in the failure of macroelements and specimen by fracture.

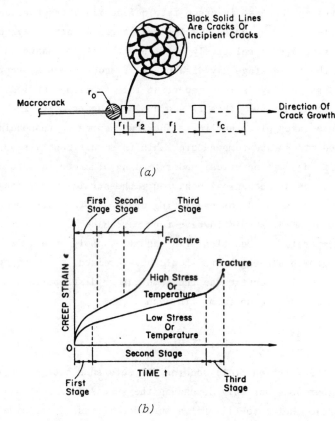

Figure 11 - Creep Failure in the Crack Tip Region
(a) creep in crack tip region
(b) creep strain versus time

Creep behaviour of metals deforming at constant load is illustrated
in Figure 11(b). The curves show that creep strain increases monotonically
with time as the material is kept in uniaxial tension at a given temperature. Only the strain due to creep is shown. The instantaneous strains
due to elastic and plastic deformation are omitted. Phenomenologically,
the total time until fracture can be divided into three stages. The first
stage is the time during which most of the transient creep takes place.
A more or less constant creep rate $d\epsilon/dt$ = const. is reached at the second
stage corresponding to the action of viscous dissipation. The third stage
involves an increase in creep rate before final fracture and this portion
is called *tertiary creep*. All three stages do not always occur. If the
final fracture is relatively brittle with little change in cross-sectional
area, the third stage may disappear. Viscous creep as represented by the
second stage plays a less important role when the applied stress and temperature are high.[7] This is shown by the upper curve in Figure 11(b) where
fracture takes place earlier. The lower curve corresponding to lower
applied stress and temperature exhibits significant viscous creep.

Although a theoretical understanding of creep is more difficult than
other types of mechanical behaviour, the strain energy density criterion
[8,9] remains valid for predicting crack growth under creep. The computation of dW/dV should involve four variables: creep strain, time, stress
and temperature. Equations (15) and (19) apply equally well to creep
crack growth although care should be exercised in computing the strain
energy density function as the temperature distribution in the crack tip
region may not be uniform. In such a case,

$$\frac{dW}{dV} = \int_{o}^{\sigma_{ij}} \sigma_{ij}\, d\epsilon_{ij} + f(\Delta T) \tag{23}$$

where the function $f(\Delta T)$ denotes the reference energy level. Failure
behaviour does not only depend on the stress and/or strain state but also
on $f(\Delta T)$. Refer to [18] for a more detailed discussion on failure prediction in nonisothermal solids. Theoretical treatment of creep fracture
remains as a topic for future investigation.

---

[7]
 At very high temperature, metals can undergo transformation to different
 crystal structures, recrystallization and grain growth. These effects
 are ordinarily excluded in creep analyses.

4. MATERIAL INHOMOGENEITY

Material inhomogeneity plays a key role in the design of structural
elements. Engineers have traditionally relied on the assumption of
material homogeneity when standardizing material properties through testing. Yield strength and fracture toughness are commonly known quantities
for characterizing polycrystalline metals. Such an approach is justified
only when load distribution and material microstructure are both sufficiently smooth and homogeneous at a specified scale level. A concentrated
load acting in a macroscopically homogeneous medium can interact with
local inhomogeneities at the microscopic scale. A system can thus be
globally homogeneous and locally inhomogeneous. The state of affairs
near the crack tip region is another example. On the other hand, composite materials such as fibre reinforced laminates are highly inhomogeneous
and their properties cannot be standardized by uniaxial tests as in the
case of metals. They must be regarded as *structures* since their behaviour
is load-specific. Homogeneity applies only to the constituents, namely
the fibre and matrix material.

4.1 Linearity and Nonlinearity

The concept of *homogeneity* is intimately associated with linearity
and nonlinearity observed on the stress and strain diagram in uniaxial
tests. Rocks or concrete-like materials are known to have initial mechanical defects or imperfections that are visible to the eye. Their stress
and strain response under uniaxial compression can be clearly divided into
three portions, Figure 12. The first portion of the curve is concave.
This corresponds to the closure of larger defects. The mean size of the
defects in the material is being reduced under compression in order to
reach a state of homogeneity characterized by linearity between stress
and strain. Within this portion of the curve, load and unload can occur
without seriously affecting the material internal structure. As the input
energy is continuously increased beyond the material's storage capacity,
the mechanism of defect coalescence takes place. This is represented by
the convex portion whose terminal point is fracture. The physical interpretation of these three portions of the stress and strain curve applies
to all materials. Since polycrystalline metals are relatively more homogeneous than rocks, the initial portion of the stress and strain curve is
always linear.

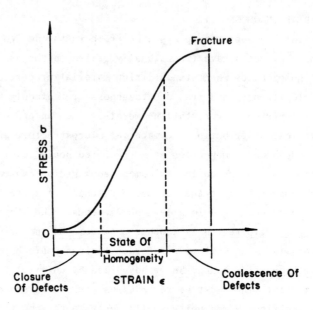

*Figure 12 - Stress and Strain Response of Rock-Like Material
          under Compression*

4.2 Interpretation of Strength

The general notion that defects or imperfections are detrimental to
the strength of solids requires scrutiny. Polycrystalline metals are
known to contain many more defects than the single crystal because of
the presence of grain boundaries. Uniaxial tensile tests have shown,
however, that polycrystalline aluminum alloys possess strength as high
as 75,000 psi while the single crystal aluminum that contains no grain
boundary defects may only have strength of 100 or 200 psi. It is then
often argued that dislocations are the cause of weakening the strength
of the single crystal. This explanation obviously does not hold because
the strength of near pure aluminum metal is only a few psi.

Strength should be associated with the degree of homogeneity governed
by load distribution and imperfection of material structure. Under uniform
loading conditions, the variation or disorder among the grains in a polycrystalline metal is less than that among the dislocations in a single
crystal. In the near pure metal, the electron movements are even more
chaotic and create an inhomogeneous state of affairs. Strength is thus
determined by the first order degree of homogeneity in the system. A single

macrocrack in an otherwise uniform solid is undesirable as it introduces local inhomogeneity because of the localized interaction of stress with material microstructure.

4.3 Energy Density Rate

The rate at which energy is dissipated within a unit volume of material can be uniquely associated with permanent disturbance or damage at the different scale levels. The nucleation and/or coalescence of pores in a single grain, the creation of microcracks in grains, and the separation of solids by macrocrack growth involve interaction of energy transfer with material microstructure. Loosely speaking, material microstructure becomes increasingly more important with decreasing energy transfer rate, say according to the order of monotonically rising load, fatigue and creep. Homogeneity, of course, also depends on the size of the volume element under observation in relation to the material microstructure. From the view of engineering application, it is essential to identify local material damage with parameters that can be measured globally.

Failure always initiates from the sites of material inhomogeneity and/or load nonuniformity. The energy per unit volume or density will, in general, fluctuate near such locations represented by the relative maxima and minima of dW/dV[8], i.e., $(dW/dV)_{max}$ and $(dW/dV)_{min}$. Their physical meanings in terms of failure prediction have already been described earlier and can be found in [8,9]. Of particular significance is the physical interpretation of the global and local stationary values[9] of dW/dV given by the pair $[(dW/dV)_{max}^{max}]_\ell$, $[(dW/dV)_{max}^{max}]_g$ and $[(dW/dV)_{min}^{max}]_\ell$, $[(dW/dV)_{min}^{max}]_g$. The former is concerned with yielding and latter with fracture. In what follows, only the case of $(dW/dV)_{min}$ will be discussed. Figure 13 shows that the locations of $[(dW/dV)_{min}^{max}]_\ell$ at L and $[(dW/dV)_{min}^{max}]_g$ at G are always separated by a finite distance ℓ along the prospective path of crack growth. The length parameter ℓ is a function of loading type and rate, specimen geometry and size, and material inhomogeneity. As the loading rate is decreased,

---

[8] The maximum of $(dW/dV)_{max}$ is associated with the site of yielding initiation and $(dW/dV)_{min}$ with fracture initiation.

[9] The local stationary values $[(dW/dV)_{max}]_\ell$ and $[(dW/dV)_{min}]_\ell$ are obtained from dW/dV by referring to system of local coordinates at the interior of the domain. The global stationary values $[(dW/dV)_{max}]_g$ and $[(dW/dV)_{min}]_g$ make reference to a fixed system of coordinates.

ℓ decreases accordingly and failure tends to become more localized and to depend more on the material microstructure. The stress and failure analysis near the crack tip can include material inhomogeneity and need not be overly complicated. A system can be divided into many subregions and only those which undergo active interaction with load require more detailed attention. Such an approach has already been practiced in analyzing crack problems where accuracy of solution is needed only in regions close to the crack point. The ℓ-concept provides a rational means of quantitatively assessing the influence of material microstructure and degree of homogeneity of a system.

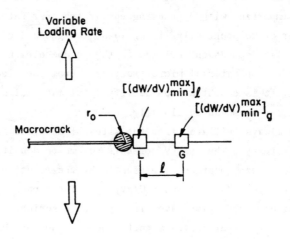

Figure 13 - Position of Local and Global Relative
Minimum Values of Strain Energy Density

5. CONCLUSIONS

The influence of material inhomogeneity in the crack tip region cannot be addressed by comparing crack dimension with microstructure alone. Local energy dissipation depends on the rate of energy transfer in a given volume of material which, in turn, defines the failure mode. It is well-known experimentally that loading rate alone can control material behaviour changing from elastic to elastic-plastic, viscoelastic or viscoplastic. Material microstructure has been damaged but not altered in its basic character. Such an influence has not been successfully included in the classical development of constitutive equations. Majority of the works relied on the assumption of homogeneity and hence are restricted to one-dimensional problems.

The establishment of threshold values of the strain energy density function with different material damage modes is a concept that can be applied at any scale level. With the advent of modern computer, the rate of material damage can be treated incrementally by performing stress and failure analysis in tandem such that the constitutive equations need not have a long memory as long as the increment of damage growth is kept sufficiently small. A pseudo-elastic model of cumulative material damage has been applied successfully to treat damage in the crack tip region of metal [19] with strain hardening and concrete with bilinear softening [20]. The accumulation of material damage in elements ahead of the crack is the dominant factor in predicting global behaviour of structural elements.

REFERENCES

1. SIH, G.C., van ELST, H.C. and BROEK, D., "Reflections on the Conference", *Prospects of Fracture Mechanics*, edited by G.C. Sih, H.C. van Elst and D. Broek, Noordhoff International Published, The Netherlands, 1974, pp. XIII-XVIII.
2. SIH, G.C., "Some Basic Problems in Fracture Mechanics and New Concepts", *Int. J. of Engineering Fracture Mechanics*, Vol. 5, No. 2, 1973, pp. 365-377.
3. SIH, G.C., "The Mechanics Aspects of Ductile Fracture", Proceedings on Continuum Models of Discrete Systems", edited by J.W. Provan, *SM Study No. 12*, Solid Mechanics Division, University of Waterloo Press, Canada, 1977, pp. 361-386.
4. SIH, G.C. and CHEN, E.P., "Effect of Material Nonhomogeneity on Crack Propagation Characteristic", *Int. J. of Engineering Fracture Mechanics*, Vol. 13, No. 3, 1980, pp. 431-438.
5. KIPP, M.E. and SIH, G.C., "The Strain Energy Density Failure Criterion Applied to Notched Elastic Solids", *Int. J. of Solids and Structures*, Vol. 11, 1975, pp. 153-173.
6. SIH, G.C., "Strain Energy Density and Surface Layer Energy for Blunted Cracks or Notches", *Stress Analysis of Notch Problems*, Vol. 5, Mechanics of Fracture, edited by G.C. Sih, Noordhoff International Publishing, The Netherlands, 1978, pp. XIII-CX.
7. OROWAN, E., "Energy Criteria of Fracture", *Welding Research Supplement*, Vol. 34, 1955, pp. 157s-160s.
8. SIH, G.C., "Experimental Fracture Mechanics: Strain Energy Density Criterion", *Experimental Evaluation of Stress Concentration and Intensity Factors*, Vol. 7, Mechanics of Fracture, edtied by G.C. Sih, Martinus Nijhoff Publishers, The Netherlands, 1981, pp. XVII-LVI.
9. SIH, G.C., "A Special Theory of Crack Propagation", *Methods of Analysis and Solutions of Crack Problems*, Vol. 1, Mechanics of Fracture, edited by G.C. Sih, Noordhoff International Publishing, The Netherlands, 1973, pp. 21-45.

10.   SIH, G.C. and MADENCI, E., "Fracture Initiation under Gross Yielding: Strain Energy Density Criterion", *J. of Engineering Fracture Mechanics* Vol. 78, No. 3, 1983, pp. 667-677.

11.   SIH, G.C. and MADENCI, E., "Crack Growth Resistance Characterized by the Strain Energy Density Function", *J. of Engineering Fracture Mechanics*, Vol. 16, (in press).

12.   SIH, G.C. and TZOU, D.Y., "Mechanics of Nonlinear Crack Growth: Effects of Specimen Size and Loading Step", *Proceedings on Modelling Problems of Crack Tip Mechanics*, edited by J.T. Pindera, Martinus Nijhoff Publishers, The Netherlands, 1984.

13.   SIH, G.C., WILLIAMS, M.L. and SWEDLOW, J.L., "Three-Dimensional Stress Distribution Near a Sharp Crack in a Plate of Finite Thickness", Air Force Materials Laboratory, Wright Patterson Air Force Base, *AFML-TR-66-242*, 1966.

14.   BENTHEM, J.P., "Three-Dimensional State of Stress at the Vertex of a Quarter-Infinite Crack in a Half Space", Delft University Technical Report, *WTHD No. 74*, September, 1975.

15.   GILLEMOT, L.F., "Criterion of Crack Initiation and Spreading", *Int. J. of Engineering Fracture Mechanics*, Vol. 8, 1976, pp. 239-253.

16.   SIH, G.C. and MOYER, E.T., Jr., "Path Dependent Nature of Fatigue Crack Growth", *Int. J. of Engineering Fracture Mechanics*, Vol. 17, No. 3, 1983, pp. 269-280.

17.   MOYER, E.T., Jr. and SIH, G.C., "Fatigue Analysis of an Edge Crack Specimen: Hysteresis Strain Energy Density", *Int. J. of Engineering Fracture Mechanics*, (in press).

18.   SIH, G.C. and MATCZYNSKI, M., "Crack Trajectories in Non-Isothermal Environments Predicted by Strain Energy Density Criterion", *J. of Theoretical and Applied Fracture Mechanics*, to appear.

19.   SIH, G.C. and MATIC, P., "A Pseudo-Linear Analysis of Yielding and Crack Growth: Strain Energy Density Criterion", *Proceedings on Defects, Fracture and Fatigue*, edited by G.C. Sih and J.W. Provan, Martinus and Nijhoff Publishers, The Netherlands, 1982, pp. 223-232.

20.   SIH, G.C. and CARPINTERI, A., "Damage Accumulation and Crack Growth in Bilinear Materials with Softening: Strain Energy Density Theory", *J. of Theoretical and Applied Fracture Mechanics*, to appear.

MODELLING PROBLEMS IN CRACK TIP MECHANICS   *Waterloo, Ontario, Canada*
CFC10, University of Waterloo   *August 24-26, 1983*

MATHEMATICAL MODELLING OF DUCTILE AND TIME-DEPENDENT FRACTURE

Michael P. WNUK
University of Wisconsin
Milwaukee, Wisconsin, U.S.A.

1. INTRODUCTION

The two papers by Griffith [1,2] provided the mathematical basis for explaining fracture in solids. Griffith bypassed the problem of an infinitely large stress, predicted at the tip of a sharp crack within the framework of the linear theory of elasticity, by employing the energy balance approach. Considering the two distinctly different states, in the thermodynamic sense, Griffith was able to discern a propagating crack from a nonpropagating crack. This "before" and "after" approach does not allow, however, to take into consideration any intermediate "activated state" which would provide a vehicle or, if you like, a mechanism to transform the "before" into the "after". In this way the detailed mechanism by which the crack opens up and begins to propagate is totally left out from the theory.

With the application of modern numerical techniques, such as the finite element method, the behaviour of cracks under general geometrical and loading conditions, can be predicted, but only if the material equation of state is known. Frequently rather arbitrary assumptions are made about the constitutive relations which describe the material response within the region immediately adjacent to the crack tip. Although a number of researchers in the field of fracture mechanics agrees that the continuum based theories of fracture should be refined by incorporating microstructural and even atomistic effects, the progress in these areas has been slow, plagued by numerous difficulties and inconsistencies. On the other hand, an impressive amount of experimental data has been collected by the physicists and metallurgists to document the events at the micro-level, such as emitting dislocations, void growth, microcrack formation, etc., which usually precede occurrence of catastrophic fracture.

Thus, there arises a need for certain plausible mathematical models which would bridge the gap between the macro and the micro fracture mechanics. Two such models, one for the description of ductile fracture, and the other aimed at explaining the time-dependent fracture resulting due to accumulation of a large number of microdefects, localized in a narrow band generated ahead of the dominant crack and measured by the so-called "damage parameter", will be discussed in this lecture.

We should point out that both approaches presented here are highly idealized, almost simplistic models of the physical reality. Nevertheless the basic assumptions used to establish the mathematical formalism appear to be in accord with the experiment, and so are the end results. This fact alone justifies the presentation which follows.

2.   MODELLING OF ELASTIC-PLASTIC FRACTURE

The primary objective of this part of the lecture is to provide a better insight into the physics of the processes involved in tearing fracture which occur at the microstructural level, and then to relate the findings to the existing continuum mechanics descriptions of fracture in elastic-plastic deformation fields. We have succeeded in suggesting certain new fracture parameters which are shown to possess a more fundamental meaning than the typical COD or J-integral representations. One important property of these new measures of material toughness within the inelastic range is their invariance to the amount of stable cracking which precedes the transition to catastrophic fracture.

Once this microscopic level of understanding fracture is established, we proceed to show that the final results do conform with the presently existing macroscopic descriptions of tearing fracture, and that an almost perfect match with the experimental data may be attained. The details of this approach are given in the recent paper by Wnuk and Mura [8].

Let us briefly outline the basic assumptions which underlie the mathematical model. One such fundamental assumption concerns the work done on a material element located a certain distance (Δ) from the crack front during the time interval preceding disintegration of the cohesive (or restraining) bond within this element. This work (say G^{Δ}), associated with an incremental crack extension, may be computed either analytically or numerically by solving a field problem which involves a moving crack. Next the requirement of invariance of G^{Δ} to the amount of crack extension leads

to the governing differential equation which predicts behaviour of a quasi-
static crack embedded in an elastic-plastic deformation field. As it
turns out the variation in the extent of the plastic zone, dR, reflecting
redistribution of the plastic strains ahead of the crack front, may be
related to an incremental crack extension, da, which has caused this redis-
bution. It may be shown that the apparent material toughness K_R associated
with a stable quasi-static crack is defined by the following functional

$$K_R = \sqrt{\frac{2}{\pi}} \left\{ \int_0^\Delta \frac{S_1(x_1)dx_1}{\sqrt{R-x_1}} + \int_\Delta^R \frac{S_2(x_1)dx_1}{\sqrt{R-x_1}} \right\} \tag{1}$$

Equation (1) results from considerations of a *structured* end-zone (see
Figure 1) associated with a quasi-static crack. It connects the extent of
plastic zone R, the process zone size Δ and the distribution of the
restraining stress $S(x_1)$. The distribution $S(x_1)$ is *not* contained within
the continuum mechanics framework and, therefore, it has to be assumed
a priori. If the end-zone is sub-divided into two sub-regions

  (1) process zone or disintegration zone, $0 < x_1 < \Delta$, and

  (2) irreversible deformation zone, $\Delta < x_1 < R$,

as shown in Figure 1, then we may visualize that while the energy adsorp-
tion process occurs within the entire end-zone, only its small fraction
(G^Δ) dissipated within the region immediately adjacent to the crack front,
has a direct effect on the failure process. In order to sustain fracture
the external loads should be applied in such a way that the energy supply
equals the so-called "essential work of fracture", $\hat{G}$. We note that while
G^Δ is a field quantity, the specific energy $\hat{G}$ is a material property. We
also note that there are two distinct parts of the restraining stress distribution. While the descending part, $S_1(x_1)$, reflects the disintegration
process occurring within the process zone, the ascending part, $S_2(x_1)$,
represents the deformation pattern at and beyond the yield point. For a
growing crack the quantities K_R and R are certain a priori unknown functions of the current crack length, a. Therefore, we have[1]

$$S_1 = S_1(x_1,\Delta,a) ; \quad S_2 = S_2(x_1,\Delta,a) ; \quad R = R(S_1,S_2\Delta,a) . \tag{2}$$

---

[1] Generalization of equation (2) onto a viscoplastic case yields:

$$S_1 = S_1(x_1,\Delta,a,\dot{a}) ; \quad S_2 = S_2(x_1,\Delta,a,\dot{a}) ; \quad R = R(S_1,S_2,\Delta,a,\dot{a}) . \tag{3}$$

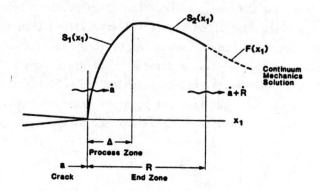

*Figure 1 - Structured Nature of the End Zone Associated
With a Quasi-Static Crack Embedded in an
Inelastic Solid. $S(x_1)$ represents the assumed
distribution of the restraining stresses.*

It is usually assumed that the process zone remains insensitive to the
amount of stable cracking, and that its size is determined solely by the
microstructural parameters. One may theorize, though, that the size of
disintegration zone, over which the restraining stress falls off from a
certain maximum level to zero, is affected not only by the material
microstructure but also by the state of stress generated in the vicinity
of the crack front. Should this be the case, then the length Δ, often
identified with the crack growth step, would also have to be considered
as a function of the current crack length, $\Delta = \Delta(a)$. For fracture in the
elastic-plastic range the current crack length is used as a time-like
variable. The case of such "variable growth step" and its effect on the
equations which describe the macroscopic (or, apparent) fracture resistance
has been examined by Wnuk and Mura, [6].

Let us now briefly review the mathematical procedures required to
obtain a resistance curve pertinent to post-yield fracture. For a quasi-
static crack propagating through a ductile solid the fundamental relation
given by Irwin

$$G_I = K_I^2/E_1 , \quad E_1 = \begin{cases} E & \text{plane stress} \\ E(1-\nu^2)^{-1} & \text{plane strain} \end{cases} \tag{4}$$

is no longer valid. It is replaced by a differential equation of this
form

$$\frac{dJ_R}{da} = f(a, J_R, J_{IC}, T_J) \; , \tag{5}$$

in which J_R is used to denote the J-integral employed as a measure of material resistance associated with stable cracking, J_{IC} is the threshold level for growth initiation, the current crack length "a" is treated as a time-like variable, and T_J denotes a dimensionless tearing modulus, as for example the one suggested by Paris, but restricted to the *initial* slope of the J_R resistance curve only. Differential equation of the type (5) results from a requirement that the amount of energy absorbed within the process zone located immediately ahead of the moving crack front is a material property, i.e.,

$$J_I^\Delta = -2 \int_0^\Delta \sigma_f(\xi) \, \frac{\partial u_y(\xi, a)}{\partial \xi} \, d\xi = \text{const} \; . \tag{6}$$

Here the symbol $u_y(\xi, a)$ denotes the displacement normal to the crack plane at the distance ξ measured from the tip of a crack of current length a, and $\sigma_f(\xi)$ denotes the "cohesive" or "fracture" stress which prevails within the structured end-zone associated with a quasi-static crack. The quantity J_I^Δ represents the separation energy release rate calculated per increment Δ of crack growth (the size of the process zone Δ is identified with a single growth-step). On the other hand, the constant appearing on the right side of equation (6) represents the "essential work of fracture", J_C^Δ or $\hat{G}$. This quantity has been determined experimentally by Cotterell [38] and by Mai [39] and shown to be invariant to the amount of crack growth.

Therefore, J_C^Δ (or $\hat{J}_C$) may be regarded to be a true measure of material toughness in a ductile solid. In contrast, the toughness parameter described by J_R-integral is not invariant during the preliminary stages of crack growth, and thus it could be labelled an "apparent" fracture toughness, as it is reflected by a typical R-curve. If a simple line-plasticity model is used to describe the deformation field near the crack tip, then the constant J_C^Δ may be replaced by the product of the effective yield stress σ_Y (which is a uniaxial yield stress augmented by an appropriate constraint factor) and the final stretch $\hat{\delta}$, i.e., the increment of the crack tip opening displacement which must be generated in a small segment

Δ of the yield zone immediately adjacent to the crack tip. The criterion
for crack propagation reads then

$$J_I^\Delta = J_C^\Delta \; ; \qquad J_C^\Delta = \sigma_Y \hat{\delta} \; . \tag{7}$$

Yet simpler form of equation (7) has been suggested by Wnuk [9] under
the name "final stretch criterion", which is obtained from the condition
(6) when $\sigma_f(\xi) = \sigma_Y$ and $J_C^\Delta = \sigma_Y \hat{\delta}$, namely

$$u_y(0, a+\Delta) - u_y(\Delta, a) = \hat{\delta}/2 \; . \tag{8}$$

If the gradient of the function $u_y(\xi, a)$, say $[\delta u_y / \delta \xi]_{\xi=\Delta}$ is known at a
small distance Δ from the crack tip (note that $\delta u_y / \delta \xi$ is singular at the
crack tip), then the expression (8) may be readily converted into a non-
linear differential equation governing a J_R-resistance curve, i.e.,

$$\frac{dJ_R}{da} = n\sigma_Y (\hat{\delta}/\Delta) - (8n\sigma_Y^2/\pi E_1) \Phi(\Delta, a) \; . \tag{9}$$

This expression represents a specific example of equation (5). The func-
tion $\Phi(\Delta, a)$ denotes the gradient $[\delta u_y / \delta \xi]_{\xi=\Delta}$, while the constant n is an
empirical numerical factor derived from the relation $J_R = 2n\sigma_Y u_y^{tip}(a)$.
The value of this material sensitive coefficient usually lies in the range
1 to 2.6. Examples of applications of the governing equation (9) for
analysis of the fully ductile fracture problems were recently given by
Smith [12]; Wnuk [13]; Wnuk and Sedmak [14].

Assuming Δ to be small versis the length of the plastic zone, Wnuk
[15] derived the following expression for the separation rate

$$J_I^\Delta = \left(\frac{8\sigma_Y^2}{\pi E_1} \right) \left\{ \Delta \frac{dR}{da} + \frac{\Delta}{2} \log \left(\frac{4eR}{\Delta} \right) \right\} , \tag{10}$$

valid within the small scale yielding range ($\Delta \le R \le a$). In the same
paper this formula was given for the energy separation rate

$$J_I^\Delta = \left(\frac{8\sigma_Y^2}{\pi E_1} \right) \left\{ \frac{a}{a+R} \frac{dR}{da} + \frac{\Delta}{2} \log \left[\frac{4eRa}{\Delta(2R+a)} \right] \right\} \tag{11}$$

valid for an arbitrary R/a ratio. This of course includes the limiting
case of the fully yielded specimen ($\Delta < R > a$), as then equation (11) can
be readily reduced to this form

$$J_I^\Delta = \left(\frac{8\sigma_Y^2}{\pi E_1}\right) \left\{ \frac{a}{R} \Delta \frac{dR}{da} + \frac{\Delta}{2} \log\left(\frac{2ea}{\Delta}\right) \right\} . \tag{12}$$

In the derivation of the last three expressions given above only the simplest geometry of a centre-cracked plate of infinite width was considered. It would be desirable to extend such analyses so that other common geometrical configurations, such as centre-cracked finite width panel, double edge tension specimen, three point bend, compact tension, and a single edge notched tension would be incorporated. In order to analyze an arbitrary geometrical configuration Wnuk [13] has suggested a somewhat more specific form of the governing equation (9). It reads as follows

$$\frac{dJ_R}{da} = n\left(\frac{\sigma_Y^2}{E_1}\right) \left\{ T_\delta - \frac{4}{\pi} \log\left[\frac{\Lambda(\Delta,a)}{\Delta}\right] \right\} . \tag{13}$$

Here, the symbol T_δ denotes the tearing modulus

$$T_\delta = \begin{cases} (E_1/\sigma_Y)(\hat\delta/\Delta) \\ \text{or} \\ (CTOA)(E_1/\sigma_Y) \end{cases} . \tag{14}$$

The acronym "CTOA" is derived from the term "crack tip opening angle", while the geometrical factor Λ depends on the given crack configuration. For a centre-cracked infinite width plate Wnuk [16] gave the factor Λ in terms of the extent of the plastic zone R and the current crack length a (with "e" denoting the base of natural logarithm) as follows

$$\Lambda(\Delta,a) = 2ea \frac{R(R+2a)}{(a+R)^2} . \tag{15}$$

Extending these considerations to a finite boundaries case Wnuk and Sedmak [14] calculated the factor Λ for a centre-cracked plate of width 2h, i.e.,

$$\Lambda(\Delta,a) = 2ea \frac{\sin^2\left[\frac{\pi(a+R)}{2h}\right] - \sin^2\left[\frac{\pi a}{2h}\right] \tan\left(\frac{\pi a}{2h}\right)}{\sin^2\left[\frac{\pi(a+R)}{2h}\right]\left(\frac{\pi a}{2h}\right)} . \tag{16}$$

This latter expression reduces to a simple form

$$\Lambda(\Delta, a) = \frac{2eh}{\pi} \sin\left(\frac{\pi a}{h}\right),$$
(16a)

when it is assumed that all of the unbroken ligament has yielded, i.e., when $a+R \cong h$.

Application of the crack growth laws (6) or (8), in which the quantity J^Δ is replaced by the forms (10) and (11), for small and large scale yielding, respectively, leads to these differential equations describing extension of a stable quasi-static crack

$$\frac{dR}{da} = \frac{\pi}{8} T_\delta - \frac{1}{2} \log(4eR/\Delta),$$
(17)

or

$$\frac{dJ_R}{da} = (4\sigma_Y^2 n/\pi E_1) \log(J_{ss}/J_R),$$
(17a)

for a contained yield situation, and

$$\frac{dR}{da} = \frac{R}{a}\left\{\frac{\pi}{8} T_\delta - \frac{1}{2} \log\left(\frac{2R^2}{ea\Delta}\right)\right\},$$
(18)

or

$$\frac{dJ_R}{da} = \left(\frac{n\sigma_Y^2}{E_1}\right)\left\{T_\delta - \frac{4}{\pi} \log\left(\frac{2ea}{\Delta}\right)\right\},$$
(18a)

for a fully yielding tensile specimen of infinite width ($\Delta \ll R \gg a$). Note that the length of the plastic zone R is used here as an alternative measure of the material resistance to cracking developed during the early stages of ductile fracture. The constants n, T_δ, σ_Y, E_1 and J_{ss} have been already explained with the exception of the last one, J_{ss}. This quantity represents the steady-state limit of the apparent fracture toughness, attained when $dJ_R/da \to o$. In other words, the J_{ss} level defines the upper plateau of the J_R-curve. It can be related to other material constants by either one of the following expressions

$$J_{ss} = \begin{cases} \left(\dfrac{2n\sigma_Y^2}{\pi e E_1}\right) \exp\left\{\dfrac{\pi}{4} T_\delta\right\}, \\[2em] \left(\dfrac{2n\sigma_Y^2}{\pi e E_1}\right) \exp\left\{\left(\dfrac{\pi E_1}{4n\sigma_Y^2}\right)(\hat{J}_C/\Delta)\right\}. \end{cases}$$
(19)

Expressions quoted here enable one to study the instabilities occurring in ductile fracture process. A nondimensional parameter, named stability index

$$\lambda = (\pi E_1/8\sigma_Y^2)\{dJ_R/da - \partial J_A/\partial a\} , \qquad (20)$$

provides a useful measure of the "degree of "stability" of any given stress state associated with a growing crack. As is seen from the formula (20) the quantity λ is positive for a stable crack, as only then the demand for the energy flow into the process zone, dJ_R/da, exceeds the actual rate of energy supply, $\partial J_A/\partial a$ (J-integral labelled with an index "A" denotes the intensity of the external field). An artist's view of the equilibrium state which exists at the tip of a quasi-static crack is shown in Figure 2.

Figure 2 - *An artist's view of the equilibrium state which exists at the crack tip. Note that the greater is the effort expended by gentleman A, who represents the external field, the stronger will be the resistance put up by gentleman R, who symbolizes the material response. For a stable crack, one would expect the demand for the energy flow into the crack tip, dJ_R/da, to exceed the available rate of energy supply, $\partial J_A/\partial a$, or shortly, $\lambda > 0$.*

3. MODELLING TIME-DEPENDENT FRACTURE IN DAMAGING MATERIALS BY THE CDM/LEFM APPROACH

3.1 Introduction

A substantial effort has been recently directed toward a more realistic description of both stages of fracture, i.e., (1) formation of the dominant crack, and (2) extension of such crack within a body containing micro-defects. The present work falls in this framework since it suggests a model for a time-dependent fracture based on two fundamental premises:

(a) the damage kinetics law of Kachanov governing spread of the damage zone ahead of the dominant crack, and

(b) a concept of simultaneous propagation of the dominant crack and the associated damage zone wherein both entities interact with each other.

It is obvious that such idealization of the creep rupture process de-rives its basic concepts from the classical mechanics of fracture, say LEFM, and from a more recent class of analyses, known under a collective name, coined by Hult in his paper co-authored with Janson [30], namely, "continuous damage mechanics", or briefly, CDM.

Since we consider damage kinetics law in its scalar form as given by Kachanov, only a one-dimensional problem of damage enlargement is treated. However, in contrast to certain earlier line-damage models, such as the one suggested by Janson and Hult [30] and then developed by Chrzanowski and Dusza [17], we incorporate (a) the effect of the finite thickness of the damage zone, and (b) the effect of interaction between the accumulation of damage and the stress concentration brought about by the existence of the dominant crack. The main crack is considered to be blunted at the very onset of its growth while the stresses ahead of the fracture front are evaluated under an assumption that the profile of the blunted crack is either elliptical or hyperbolic.

Mathematics of the problem is reduced to a numerical procedure which allows for an effective solution of the integro-differential equation which governs extension of cracks in materials capable of sustaining creep strains. To illustrate the effects of load, crack, dimension, and various material parameters which enter into the constitutive (creep) and the damage kinetics laws, certain specific geometries such as single-edge notch, double-edge notch, three-point bend and four-point bend are chosen.

3.2 Damage Band and Its Interaction with the Dominant Crack

Let us consider a one-dimensional damage zone propagating ahead of
the dominant crack and interacting with the state of stress induced by the
approaching front of fracture. According to this model, the damage zone
forms a band of width 2ρ (ρ denotes also the radius of the *blunted* crack
tip) within which the material defects of various kinds, such as microcracks or microvoids, are localized. We assume that the process of
damage localization and its subsequent propagation is influenced by the
advance of a crack and, vice-versa, the rate of spreading crack is affected
by an increasing density of micro-defects accumulated within the damage zone.
We consider a quasi-static process of crack extension so that most of the
quantities involved are time-dependent, including the internal damage
parameter, $\omega = \omega(t)$, which varies between the limits $\omega = 0$ and $\omega = 1$.
While the lower limit of ω corresponds to an undamaged material, the upper
limit is tantamount to an attainment of the critical state identified here
with a local collapse occurring at a material point where $\omega \to 1$.

To follow the history of deformation at a generic point P located a
critical distance Δ from the tip of a sharp crack, or placed directly on
the elliptical (or hyperbolic) front of a blunted crack, i.e., at the distance $r = \rho/2$ from the focus of the ellipse as shown in Figure 3, we shall
consider the sequence of stress states induced at this point by a moving
dominant crack. The stress at P due to an approaching crack whose length
varies between a_o and a can be derived from a well-known LEFM formulae
(compare Creager and Paris [31]):

$$\sigma_x = - \frac{K_I}{\sqrt{2\pi r}} \frac{\rho}{2r} \cos \frac{3\theta}{2} + \frac{K_I}{\sqrt{2\pi r}} \cos \frac{\theta}{2} \left[1 - \sin \frac{\theta}{2} \sin \frac{3\theta}{2}\right] + \cdots ,$$

$$\sigma_y = + \frac{K_I}{\sqrt{2\pi r}} \frac{\rho}{2r} \cos \frac{3\theta}{2} + \frac{K_I}{\sqrt{2\pi r}} \cos \frac{\theta}{2} \left[1 + \sin \frac{\theta}{2} \sin \frac{3\theta}{2}\right] + \cdots , \quad (21)$$

$$\tau_{xy} = - \frac{K_I}{\sqrt{2\pi r}} \frac{\rho}{2r} \sin \frac{3\theta}{2} + \frac{K_I}{\sqrt{2\pi r}} \sin \frac{\theta}{2} \cos \frac{\theta}{2} \cos \frac{3\theta}{2} + \cdots .$$

In what follows we shall assume that only the component σ_y is responsible for cleavage fracture. This may not be an entirely realistic assumption, but it certainly is the simplest and most justifiable for a quasi-
brittle solid. Next, it is easily shown that at the crack periphery that
is at $\theta = 0$ and $r = \rho/2$, the stress σ_y reduces to

$$[\sigma_y]_{\substack{crack \\ front}} = \frac{2\ K_I}{\sqrt{\pi\rho}} \quad \text{(blunted crack)} \tag{22}$$

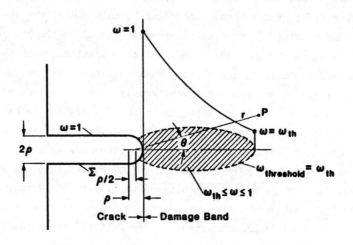

Figure 3 - Crack of finite width and the associated damage band. Note that the crack front (Σ) is defined as a locus of points at which ω = 1.

If all the *preceding* states associated with the current crack length ($a_o < a' < a$) are to be accounted for, then the formula (22) has to be replaced by (see Figure 4):

$$[\sigma_y]_\Sigma = [\sigma_y]_{r=a-a'+\rho/2} = \frac{K_I(a')}{\sqrt{2\pi}}\ \frac{\rho+a-a'}{\left(\frac{1}{2}\ \rho+a-a'\right)^{3/2}}\ . \tag{23}$$

We employ the symbol Σ to denote the boundary of the damage zone at which the internal damage parameter attains the critical level, $\omega \rightarrow 1$.

Note that the current crack length a' is treated here as a time-like parameter which, during the propagation phase, is related uniquely to time t', through the obvious relation

$$dt' = \frac{da'}{\dot{a}(a')}\ , \qquad t_1 \le t' \le t\ . \tag{24}$$

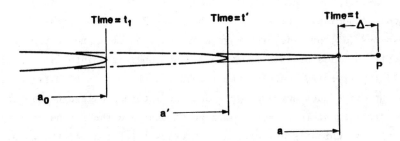

Figure 4 - Stages of Crack Advancement Toward the Control Point:
  (a) *initiation of growth at t_1,*
  (b) *location of the crack front at an intermediate instant t', and*
  (c) *collapse occurring at the control point at time t.*

We note that the lower limit t_1 corresponds to the initiation of crack extension (or, equivalently, to the end of the incubation period during which the crack is dormant), while the upper limit t marks the instant at which the material element located at the generic point P collapses.

According to Kachanov's hypothesis, the rate of damage accumulation $d\omega/dt$ is proportional to a certain power of the stress $[\sigma_y]_\Sigma$, namely

$$\frac{d\omega}{dt} = C \left[\frac{\sigma_\Sigma}{1-\omega}\right]^\nu \tag{25}$$

Here ω is a certain function of time, σ_Σ is given by equation (23), while C and ν are material constants. In what follows we shall consider an integral of expression (25), namely

$$\int_o^\omega (1-\omega')^\nu d\omega' = C \int_o^{t'} \sigma_\Sigma^\nu (t'')dt'' . \tag{26}$$

When the internal damage parameter ω attains one, the upper limit in the integral on the right of equation (26) approaches the instant t which marks the final collapse of a material element at P. Therefore, we have

$$\Omega_c = \int_o^1 (1-\omega)^\nu d\omega' = C \int_o^t \sigma_\Sigma^\nu (t')dt' . \tag{27}$$

If we consider the nondimensional damage parameter Ω_c as a measure of the critical accumulation of strain at the point P, then the first of the expressions (27) provides the governing equation of a propagating crack. The left hand side of this equation may be subdivided into two parts: the first part corresponds to the incubation phase $(0 \leq t' \leq t_1)$ during which the crack does not propagate but the internal damage parameter ahead of the crack front increases from zero to one; and the second part $(t_1 \leq t' \leq t)$ which is associated with crack extension and the ensuing interaction between the damage accumulation (ω) and the stress induced by an advancing crack. The instant t denotes the termination of the second phase of fracture coinciding with collapse of the material element located at the point P. Therefore, we rewrite equation (27) as follows:

$$C \int_0^{t_1} \sigma_\Sigma^\nu(t')dt' + C \int_{t_1}^{t} \sigma_\Sigma^\nu(t')dt' = \Omega_c \ . \tag{28}$$

Since the first of the integrals appearing above reduces to

$$\Omega_c \ \rho^{\nu/2} \left[\frac{\rho+a-a_0}{(\rho+2a-2a_0)^{3/2}} \right] \ ,$$

our criterion for time-dependent failure, which governs propagation of the dominant crack, assumes this form:

$$C \int_{t_1}^{t} \sigma_\Sigma^\nu(t')dt' = \Omega_c \left[1 - \left(\frac{\sqrt{\rho}(\rho+a-a_0)}{(\rho+2a-2a_0)^{3/2}} \right)^\nu \right] \ . \tag{29}$$

If now expression (23) is substituted for σ_Σ and the time integral is written as an integral over the prior crack lengths, see equation (24), then we have

$$C \int_{a_0}^{a} \left[\frac{K_I(a')(\rho+a-a')}{\sqrt{2\pi} \ (\rho/2+a-a')^{3/2}} \right]^\nu \frac{da'}{\dot{a}(a')} = \Omega_c \left[1-\rho^{\nu/2} \left(\frac{\rho+a-a_0}{(\rho+2a-2a_0)^{3/2}} \right)^\nu \right] \ . \tag{30}$$

This constitutes an integral equation for the unknown growth rate, $\dot{a}$, as a function of the crack length a. It is a linear Volterra type equation in which the unknown function is the reciprocal of the crack growth rate, namely $1/\dot{a}$. Although it is possible to construct a closed form analytical solution for this equation by Laplace transformation, we shall employ a numerical technique which leads to a relatively uncomplicated and an efficient algorithm. This approach permits quick and straightforward generation of the output data pertaining to creep rupture by using a computer. The numerical procedures involved are omitted in this brief text, and only the final results are summarized.

The equation of a moving crack, equation (30), which was obtained by combining the stress distribution in a brittle solid induced by a blunted crack and the Kachanov law of damage accumulation, may be subsequently reduced to an expression for the *rate* of the growing crack, i.e.,

$$\frac{da}{dt} = \frac{C\,\sigma_{\Sigma}^{\nu}(a)}{C\left(\dfrac{\nu}{2\rho}\right) \displaystyle\int_{a_0}^{a} \dfrac{\sigma_{\Sigma}^{\nu}(a')}{\lambda(a')} \dfrac{\frac{3}{2} + \lambda(a')}{\frac{1}{2} + \lambda(a')} \dfrac{da'}{\dot{a}(a')} + f_1(a)}\,, \tag{31}$$

in which

$$\sigma_{\Sigma}(a) = 2K_I(a)/\sqrt{\pi\rho}\,,$$

$$\sigma_{\Sigma}(a') = \frac{K_I(a')}{\sqrt{2\pi\rho}} \frac{\frac{1}{2} + \lambda(a')}{\lambda^{3/2}(a')}\,,$$

$$f_1(a) = \Omega_c\left(\frac{\nu}{\rho}\right) \frac{\left(\frac{3}{2} + \lambda_0\right)\left(\frac{1}{2} + \lambda_0\right)^{\nu-1}}{(2\lambda_0)^{\frac{3}{2}\nu + 1}}\,, \tag{31a}$$

$$\lambda = \frac{1}{2} + \frac{a}{\rho} - \frac{a'}{\rho}\,, \qquad \lambda_0 = \frac{1}{2} + \frac{a}{\rho} - \frac{a_0}{\rho}\,,$$

This form is obtained from equation (30) by the differentiation of both sides of equation (30) with respect to a. Equation (31) predicts the rate of crack growth as a certain function of the crack length, a, treated here as a time-like variable. In particular, the initial rate of crack extension, evaluated at the instant of termination of the incubation phase of creep rupture, i.e., when $t = t_1$ and $a = a_0$, is predicted from (31) as follows:

$$\left(\frac{da}{dt}\right)_{ini} = \frac{2^{\nu}C\rho}{2\nu\Omega_c}\left[\left(\frac{K_I}{\pi\rho}\right)\right]_{a=a_0}^{\nu/2}\,, \qquad \text{at } t = t_1\,. \tag{32}$$

It turns out that the parameters Ω_c and t_1 are related by a simple formula

$$t_1 = \frac{\Omega_c}{C} \frac{(\pi\rho)^{\nu/2}}{2^\nu K_I^\nu(a_0)} . \tag{33}$$

Numerical integration of equation (31) allows for determination of the function $\dot{a} = f(a)$. This has been accomplished for four different pre-cracked specimen configurations, namely: (a) single-edge notch, (b) double edge notch, (c) three-point bend, and (d) four-point bend. In all instance it has been shown that in a constant load test the growth rate *decreases* initially to a minimum which is only slightly lower than the starting rate given by equation (32), and then it increases rapidly to a certain final (sometimes unbounded level. This rapid increase in the rate of creeping crack is interpreted as the global failure of the specimen. If we denote the duration of the incubation phase (i.e., the dormant crack) by t_1 and the time of final rupture by t_*, then it is of ultimate interest to pre-dict the numerical values of times t_1 and t_* for any specific geometry of the structural element, the type of loading (tension or bending) and the crack configuration involved (single- or double-edge notch). This has been done by direct integration of the governing equation (31), see Figure 5.

We may conclude, therefore, that the description of creep crack propa-gation based on consideration of interacting macro- and micro-defects (or, a "cross" between LEFM and CDM) leads to a more realistic mathematical model. The approach followed here provides a greater flexibility in data reduction and interpretation as compared to the standard CDM model. This is due to the incorporation of two distinct sets of parameters governing the rate of crack growth and time to failure for any specified load/crack configuration. The first of the two sets corresponds to the quantities used in the macroscopic or field theory of fracture, such as the stress intensity factor K_I and the material toughness K_{IC}, while the second result from the microstructural considerations underlying the laws which govern formation and spread of micro-defects in a stressed solid.

We realize that further refinements of the model are necessary. In particular, it is desirable that certain novel constitutive relations and damage growth laws are developed to better represent the behaviour of various types of materials. One such possibility is offered by the recent work of Krajcinovic [34,35].

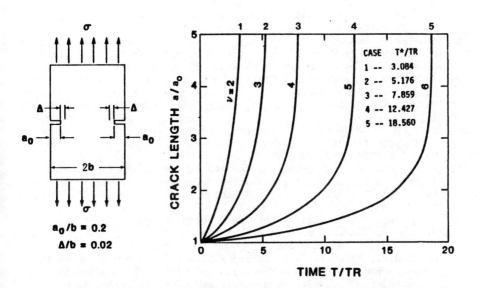

Figure 5 - *Propagation of crack due to damage accumulation. Figure shows the effect of the exponent ν, appearing in the Kachanov law, $\dot{\omega} = C[\sigma_{\Sigma}/(1-\omega)]^{\nu}$, on the time to failure T^*. The reference TR is chosen as the time to initiation of crack growth (t_1 in the text). Geometry of the fracture specimen is that of a double-edge notch panel in tension, while the load level was assumed to equal the yield stress. The curves shown resulted from a numerical integration of equation (31).*

ACKNOWLEDGEMENTS

This research constitutes part of a program supported by the Office of Naval Research under Contract No. N00014-81-K0215.

REFERENCES

1. GRIFFITH, A.A., "The Phenomena of Rupture and Flow in Solids", *Phil. Trans. Royal Soc., London,* Vol. A221, 1921, pp. 163-198.

2. GRIFFITH, A.A., "The Theory of Rupture", *Proc. of the First Int. Congress of Applied Mechanics,* Delft, The Netherlands, 1924.

3. WNUK, M.P., "Stable Phase of Ductile Fracture in Two- and Three-Dimensions", *Trans. ASME, J. of Applied Mechanics,* Vol. 48, 1981, pp. 500-508.

4. WNUK, M.P. and MURA, T., "Comparative Study of Models for Tensile Quasi-Static Fracture", *Int. J. Eng. Sci.,* Vol. 19, 1981, pp. 1517-1527.

5. WNUK, M.P., "Discontinuous Extension of Fracture in Elastic-Plastic Deformation Field", *Proc. of the Int. Sym. on Elastic-Plastic Fracture,* ASTM STP 803, 1982, pp. I-159 - I-175.

6. WNUK, M.P. and MURA, T., "Extension of a Stable Crack at a Variable Growth Step", presented at *ASTM 14th Nat. Sym. on Fracture,* ASTM STP 791, 1982, pp. I-130 - I-158.

7. WNUK, M.P., "Instability Problems in Ductile Fracture", *Proc. of ICF5*, Vol. 3, edited by D. Francois, Pergamon Press, 1981.

8. WNUK, M.P. and MURA, T., "Effect of Microstructure on the Upper and Lower Limits of Material Toughness in Elastic-Plastic Fracture", *J. of Mechanics of Materials*, Vol. 2, 1983, pp. 33-46.

9. WNUK, M.P., "Accelerating Crack in a Viscoelastic Solid Subject to Subcritical Stress Intensity", *Proc. of Int. Conf. on Dynamic Crack Propagation*, edited by G.C. Sih, Noordhoff, Leyden, The Netherlands, 1972.

10. WNUK, M.P., "Quasi-Static Extension of a Tensile Crack Contained in a Viscoelastic-Plastic Solid", *Trans. ASME, J. Applied Mechanics*, Vol. 41, 1974, pp. 234-242.

11. SMITH, E., "Some Observations on Viability of Crack Tip Opening Angle as a Characterizing Parameter for Plane Strain Crack Growth in Ductile Materials", *Int. J. Fracture Mechanics*, Vol. 17, 1980, pp. 443-448.

12. SMITH, E., "The Invariance of the J versus Δc Curve for Plane Strain Crack Growth in Ductile Materials", *Int. J. Fracture Mechanics*, Vol. 17, 1980, pp. 373-380.

13. WNUK, M.P., "Stability of Tearing Fracture", lecture given at the Summer School on Fracture Mechanics and Engineering Applications, sponsored by University of Belgrade and GOSA, Belgrade, 1980.

14. WNUK, M.P. and SEDMAK, S., "Final Stretch Model of Stable Fracture", presented at *13th ASTM National Sym. on Fracture Mechanics*, ASTM STP 743, 1980, pp. 500-508.

15. WNUK, M.P., "Transition from Slow to Fact Fracture under Post-Yield Condition", *Proc. of ICM3*, Vol. 3, Cambridge, Pergamon Press, 1979, pp. 549-561.

16. WNUK, M.P., "Occurrence of Catastrophic Fracture in Fully Yielded Components: Stability Analysis", *Int. J. Fracture Mechanics*, Vol. 15, 1979, pp. 553-581.

17. CHRZANOWSKI, M. and DUSZA, E., "Creep Crack Propagation in a Notched Strip", *J. Mecanique Appliquee*, Vol. 4, 1980, pp. 461-474.

18. RIEDEL, H. and RICE, J.R., *Proc. of ASTM 12th National Symposium on Fracture Mechanics*, edited by P.C. Paris, ASTM STP 700, 1980, p. 112.

19. RIEDEL, H. and WAGNER, W., "The Growth of Macroscopic Cracks in Creeping Materials", *Proc. of ICF5*, Vol. 3, edited by D. Francois, Pergamon Press, 1981, pp. 683-690.

20. HUI, C.Y. and RIEDEL, H., *Int. J. Fracture Mechanics*, Vol. 17, 1981.

21. RIEDEL, H., "Creep Deformation at Crack Tips in Elastic-Viscoplastic Solids", *J. Mech. Phys. Solids*, Vol. 29, 1981, pp. 35-49.

22. KNOTT, J.F., "Micromechanisms of Fibrous Crack Extension in Engineering Alloys", *Met. Sci.*, Vol. 14, 1980, p. 327.

23. GOOCH, D.J., "The Effect of Microstructure of Creep Crack Growth in Notched Bend Tests in 0.5 Cr-0.5 Mo 0.75V Steel", *Mat. Sci. Eng.*, Vol. 27, 1977, pp. 57-68.

24. WILKINSON, D.S., THYGARAJAN, N., ABIKO, K.A. and POPE, D.P., "Compositional Effects on the Creep Ductility of a Low Alloy Steel", *Met. Trans.*, Vol. IIA, 1980, p. 1827.

25. KACHANOV, L.M., "Time of the Rupture Process under Creep Conditions", (in Russian), *Izv., Akad., Nauk S.S.S.R.*, No. 8, 1958, pp. 26-31.

26. KACHANOV, L.M., *Foundations of Fracture Mechanics*, (in Russian), Nauka, Moscow, 1974.

27. KACHANOV, L.M., "Some Problems of Creep Fracture Theory", *Advances in Creep Design*, Appl. Science Publishers, London, 1971, pp. 21-29.

28. HULT, J., "Creep in Continua and Structures", *Topics in Applied Continuum Mechanics*, Springer, Vienna, 1974, pp. 137-155.

29. KACHANOV, L.M., "Crack and Damage Growth in Creep. A Combined Approach", *Int. J. Frachture Mechanics*, Vol. 16, 1980, pp. R179-181.

30. JANSON, J. and HULT, J., "Fracture Mechanics and Damage Mechanics; A Combined Approach", *J. Mecanique Appliquee*, No. 1, 1977, pp. 69-84.

31. CREAGER, M. and PARIS, P.C., "Elastic Field Equations for Blunt Cracks with Reference to Stress Corrosion Cracking", *Int. J. Fracture Mechanics*, Vol. 3, 1967, pp. 247-252.

32. BROBERG, H. and WESTLUND, R., "Creep Rupture of Specimens with Random Material Properties", *Int. J. Solids and Structures*, Vol. 14, 1978, pp. 959-970.

33. ROUSSELIER, G., "Finite Deformation Constitutive Relation Including Ductile Fracture Damage", *Proc. of the IUTAM Sym. on 3D Constitutive Relations and Ductile Fracture*, edited by S. Nemat-Nasser, North-Holland Publishing Co., 1981.

34. KRAJCINOVIC, D. and FONSEKA, G.U., "Continuous Damage Theory of Brittle Materials", (Parts I and II), *J. Appl. Mech.*, Vol. 48, 1981, p. 809.

35. KRAJCINOVIC, D. and SILVA, M.A.G., "Statistical Aspects of Continuous Damage Theory", *Int. J. Solids and Structures*, Vol. 18, 1982, p. 551.

36. SALGANIK, P.L., "Mechanics of Body Containing a Large Number of Cracks", (in Russian), *Mekh. Tverdogo Tela*, No. 4, pp. 149-159.

37. VITEK, V., "Diffusional Growth of Intergranular Cavities in a Uniform Stress Field and Ahead of a Crack-Like Stress Concentrator", *Met. Sci.*, Vol. 14, 1980, pp. 403-407.

38. COTTERELL, B., "Plane Stress Ductile Fracture", *Proc. of Int. Conf. on Fracture Mechanics and Technology*, Hong Kong, 1977, Vol. II, Sijthoff and Noordhoff Int. Publishers, The Netherlands, 1977, pp. 785-795.

39. MAI, Y.W., private communication, 1979.

Part 2
Particular Basic Problems Session
(Invited)

MODELLING PROBLEMS IN CRACK TIP MECHANICS *Waterloo, Ontario, Canada*
CFC10, University of Waterloo *August 24-26, 1983*

NONLOCAL STRESS FIELDS OF DISLOCATIONS AND CRACK

A. Cemal ERINGEN
Princeton University
Princeton, New Jersey, U.S.A.

1. INTRODUCTION

The state of stress in the core region around the tip of a sharp crack is of crucial importance in the discussion of crack instability and fracture. Dislocations that are issued from the crack tip (or present in the solid), affect the stress field appreciably depending on their locations and distributions. It is well-known that the classical elasticity solutions of the problem of crack and dislocation interaction fail in the core region since they contain stress singularities both at the crack tip and at dislocations. In this region, one must either resort to atomic lattice models or employ the nonlocal continuum theory.

In several previous papers [1-3], I have shown that the nonlocal elasticity solution of crack and dislocation problems do not contain singularities. Moreover, the fracture toughness and theoretical strength predicted by the theory are in close agreement with atomic results and experiments.

In this paper, I present the solution of the field equations of nonlocal elasticity for:

(i) a single screw dislocation (Section 3),

(ii) uniformly distributed screws (Section 3),

(iii) a line-crack subject to a constant anti-plane shear (Section 4),

(iv) a single edge dislocation (Section 5), and

(v) some results on the interactions of screw dislocations with an anti-plane crack (Section 6).

Displacement and stress fields are calculated throughout the core region and beyond. It is shown that the stress field vanishes at the crack tip, increasing to a maximum near the tip when the dislocations are absent. At larger distances from the crack tip, the shear stress approaches classical elasticity values.

By equating the maximum shear stress to the yield stress of the crystal, we arrive at a natural fracture criterion. Theoretical strengths estimated this way for various modes are in agreement with estimates based on the atomic theory.

It is well known that when a crystal contains large number dislocations, the theoretical strength is reduced considerably. By considering a line distribution of screw or edge dislocations, I give formulas for this reduction (Sections 3 and 6). The case of line crack in Mode III is solved exactly (Section 4). The stress intensity factors calculated agree well with known experimental results.

In Section 6, the state of stress in a perfect crystal containing a crack and a screw dislocation is discussed. The reduction is noted in the stress intensity factor with the distance from the crack tip of a single screw, especially when the dislocation is very near the crack tip. When the dislocations are located at some distance from the crack tip, it is permissible to superpose the stress fields of crack and dislocations. In this way, the reduction in the stress intensity factor is found to depend on the distribution of dislocations. For a uniform distribution of dislocations, the reduction in the stress intensity factor is shown to depend on the total number of dislocations. An equation is given for this reduction.

Based on these results and other theoretical work on the subject, e.g., [4,5,6] I am of the opinion that for the ultimate understanding of the crack tip problem and consequently mechanics of fracture, a proper tool is the recently developed nonlocal continuum theory. Of course, as in all scientific disciplines, the knowledge available from the atomic lattice dynamics for various ideal cases and some crucial experiments must always be used to supplement the theory and/or to provide acid tests for the theoretical work.

2. BASIC EQUATIONS

For linear homogeneous and isotropic nonlocal elasticity, the field equations are embodied in the set of equations, [1,7,8]:

$$t_{k\ell,k} + \rho(f_\ell - \ddot{u}_\ell) = 0 , \tag{2.1}$$

$$t_{k\ell} = \int_V \alpha(|\underset{\sim}{x}' - \underset{\sim}{x}|)\sigma_{k\ell}(\underset{\sim}{x}')dv(\underset{\sim}{x}') , \tag{2.2}$$

$$\sigma_{k\ell} = \lambda e_{rr}\delta_{k\ell} + 2\mu e_{k\ell} , \tag{2.3}$$

$$e_{k\ell} = \frac{1}{2} (u_{k,\ell} + u_{\ell,k}) \tag{2.4}$$

where $t_{k\ell}$, ρ, f_ℓ and u_ℓ are, respectively, the stress tensor, mass density, body force density and the displacement vector. λ and μ are the Lamé constants and α is the *attenuation function* which depends on the distance $|\underset{\sim}{x}'-\underset{\sim}{x}|$. Equations (2.1), (2.3) and (2.4) are those known from classical elasticity, but equation (2.2) replaces the Hooke's law. According to (2.2), the stress at ɛ reference point $\underset{\sim}{x}$ in the body depends on strains $e_{k\ell}(\underset{\sim}{x}',t)$ at *all points* $\underset{\sim}{x}'$ of the body. This dependence is alleviated with the attenuation function α. α has the dimension of length^{-3} in three-dimensional space, so that it depends on an internal characteristic length scale ε. We take this length scale to be proportional to the lattice parameter a for pure crystals, i.e., we write

$$\varepsilon = e_0 a = \tau\ell , \tag{2.5}$$

where e_0 is a constant appropriate to each material. Alternatively, one can use an external characteristic length ℓ with a constant of proportionality τ. The internal characteristic length depends on the internal structure of the solid and sensitivity of the physical phenomena to be investigated. For example, for granular materials, it may be selected as the mean grain size or distance and for composite materials, the average fibre distance, etc.

In a previous paper [5], I have shown that several convenient forms exist for $\alpha(|\underset{\sim}{x}|)$. For a two-dimensional lattice structure with $e_0 = 0.39$,

$$\alpha(|\underset{\sim}{x}|) = (2\pi\varepsilon)^{-1} K_0(\sqrt{\underset{\sim}{x}\cdot\underset{\sim}{x}}/\varepsilon) \tag{2.6}$$

provides an excellent match of dispersion curves of plane waves based on the nonlocal theory and the atomic lattice dynamics. The maximum error is less than 6 percent within the entire Brillouin zone. Moreover, equation (2.6) is a delta sequence so that as $\varepsilon \to 0$, $\alpha \to \delta(|\underset{\sim}{x}|)$ and (2.2) reverts to Hooke's law of classical elasticity. In fact, we note that α satisfies the following partial differential equation

$$(1-\varepsilon^2\nabla^2)\alpha = \delta(|\underset{\sim}{x}'-\underset{\sim}{x}|) . \tag{2.7}$$

Consequently, if we apply this operator to (2.2), we obtain

$$(1-\varepsilon^2\nabla^2)t_{k\ell} = \sigma_{k\ell} \ . \tag{2.8}$$

Divergence of (2.8), upon using (2.1), (2.3) and (2.4), leads to

$$(\lambda+\mu)u_{k,k\ell}+\mu u_{\ell,kk} + (1-\varepsilon^2\nabla^2)(\rho f_\ell-\rho\ddot{u}_\ell) = 0 \ . \tag{2.9}$$

For the static case with vanishing body forces, this gives Navier's equations

$$(\lambda+\mu)u_{k,k\ell} + \mu u_{\ell,kk} = 0 \ . \tag{2.10}$$

Note, however, that the stress tensor is not $\sigma_{k\ell}$ but $t_{k\ell}$, and it requires solving (2.8) to determine $t_{k\ell}$.

3. SCREW DISLOCATION

For the screw dislocation, the displacement field is the same as in classical (local) theory of elasticity [9]. In terms of the polar coordinates r, θ, z, it is given by

$$u_z = \frac{b}{2\pi}\,\theta \ , \qquad\qquad u_r = u_\theta = 0 \ , \tag{3.1}$$

where b is the Burger's vector, Figure 1. Classical stress fields are given by

$$\sigma_{\theta z} = b/2\pi r \ , \qquad \text{all other } \sigma_{k\ell} = 0 \ . \tag{3.2}$$

The nonlocal stress field is determined by solving (2.8), i.e.,

$$(1-\varepsilon^2\nabla^2)t = \sigma \ , \tag{3.3}$$

where

$$t \equiv t_{23}- it_{13} \ , \qquad \sigma = \sigma_{23}-i\sigma_{13} = \frac{\mu b}{2\pi r}\,e^{i\theta} \ , \tag{3.4}$$

$$t_{\theta z} - it_{rz} = te^{-i\theta} \ . \tag{3.5}$$

The general solution of (3.3) is given by

$$t = \sigma+K_1(r/\varepsilon)(Ae^{i\theta}+Be^{-i\theta}) \tag{3.6}$$

where A and B are constants and $K_1(\rho)$ is the modified Bessel's function.

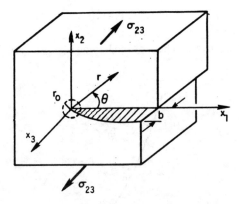

Figure 1 - Screw Dislocation

We imagine the centre of the dislocation as a limit of a small circular arc with radius $r = r_0$. On this free surface, t_{rz} must vanish as $r_0 \to 0$. Using (3.5) and (3.6), this edge condition will be fulfilled if $B = 0$ and $A = -\mu b/2\pi\varepsilon$. Consequently, we have [5]

$$T_\theta(\rho) \equiv (2\pi\varepsilon/\mu b) t_{\theta z} = \rho^{-1}[1-\rho K_1(\rho)] \ , \qquad t_{rz} = 0 \ , \tag{3.7}$$

where

$$\rho \equiv r/\varepsilon \ . \tag{3.8}$$

The stress field (3.7) has no singularity at $\rho = 0$. In fact, $t_{\theta z}$ vanishes at the centre of the dislocation. This is in contradiction to the classical elasticity solution which gives infinite stress at $r = 0$. In Figure 2, we display $T_\theta(\rho)$ as a function of ρ. The maximum stress occurs at $\rho = 1.1$ and is given by

$$T_{\theta max} = 0.3993 \ , \qquad \rho_c = 1.1 \ . \tag{3.9}$$

It is natural to assume that when $t_{\theta z max}$ becomes equal to the theoretical strength t_y^c, the solid will rupture. Thus,

$$t_y^c/\mu = 0.3993 \ \frac{b}{2\pi\varepsilon} \ . \tag{3.10}$$

If we write $h = \varepsilon/0.3993$, this agrees with the estimate of Frenkel, based on the atomic model (Kelly [10], p. 12). Using $\varepsilon = e_0 a = 0.39a$ for pure aluminum crystal, (3.10) gives

$$t_y^c/\mu = 0.12 , \qquad \{A\ell: [111]<1\bar{1}0>\} , \qquad (3.11)$$

which is very close to the theoretical result $t_y/\mu = 0.11$, based on atomic models.

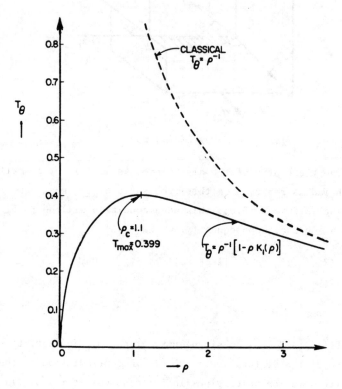

Figure 2 - Non-Dimensional Hoop Stress for Screw Dislocation

Figure 2 also shows that the maximum stress occurs at $\rho = \rho_c$, not at the centre of the dislocation. This implies that the fracture is initiated at $\rho = \rho_c$, *not* at $\rho = 0$ and there is a low stress region, $0 < \rho < \rho_c$ within the core.

Suppose that the solid contains a dislocation distribution with density $n(x)$ over an interval $|x| \leq d$, $y = 0$. The shear stress along the x-axis is obtained by

$$t_{\theta z} = \frac{\mu b}{2\pi} \int_{-d}^{d} \text{sgn}(x-x_0) \left[\frac{1}{x-x_0} - \frac{1}{\varepsilon} K_1(|x-x_0|/\varepsilon) \right] n(x_0) dx_0 . \qquad (3.12)$$

Given $t_{\theta z}$, this equation can be solved for $n(x)$, which determines the distribution of screws maintained by the stress field. Classical elasticity solution of (3.12) (where K_1 is absent) for $t_{\theta z}$ = const. is well-known ([9], p. 700). Here we examine a simpler situation, namely given $n(x)$ determined the stress field. To this end, we take $n = n_0$ = constant, which leads to

$$t_{\theta z}/t_d = \ln\left|\frac{\xi+1}{\xi-1}\right| + K_0(d_0|\xi+1|) - K_0(d_0|\xi-1|) , \qquad (3.13)$$

where

$$t_d = \mu b n_0/2\pi , \qquad \xi = x/d , \qquad d_0 = d/\varepsilon = d/e_0 a . \qquad (3.14)$$

The distribution of the shear stress (3.13) as a function of ξ is shown in Figure 3, for $d_0 \geq 5$. Behaviour of $t_{\theta z}$ is governed, basically, by the first term in (3.13), except near $\xi = 1$. For example, for $d_0 = 5$, $K_0(5) \simeq 3.7 \times 10^{-3}$, so that the second term in (3.13) can be neglected for $\xi \geq 0$. At $\xi = 1$, we have

$$t_{\theta z max} = (\mu b N/2\pi d)\ln(d/e_0 a) , \qquad (3.15)$$

where we also wrote $n_0 = N/d$, N being the total number of dislocations over a distance d. Combining (3.10) with (3.15), we obtain

$$N = 0.3993 \frac{t_{\theta z}}{t_y^c} \frac{d_0}{\ln d_0} . \qquad (3.16)$$

When $t_{\theta z} = t_y^c$, yielding begins and at this stress level, we have

$$N_{max} = 0.3993 \frac{d_0}{\ln d_0} . \qquad (3.17)$$

For $d_0 = e \simeq 2.73$, corresponding to $d = 2.73 e_0 a$, we have the lowest number $N_{max} = 1.09$. This is approximately one dislocation over a half lattice parameter. With the increase of d, N_{max} increases. For $d = 4.5 \times 10^{-4}$ cm for Fe.crystal $d_0 = 4.02 \times 10^4$, and we obtain $N_{max} = 1514$, which may be conservative, since the density function $n(x)$ is not constant for $t_{\theta z} = t_y^c$ throughout the interval $|x| < d$, $y = 0$.

A second interpretation of (3.15) may be made if we consider $B = 2dN$ as Burger's vector of a large screw and write $t^c/\mu = 0.3993 B/2\pi\varepsilon$, in accordance with (3.10). At yield $t_{\theta z max} = t_y^d$, where t_y^d is the yield stress when there is a dislocation distribution in the solid. With this interpretation, (3.15) gives

$$t_y^d/t_y^c \simeq 1.25 \, \frac{\ln d_0}{d_0} . \tag{3.18}$$

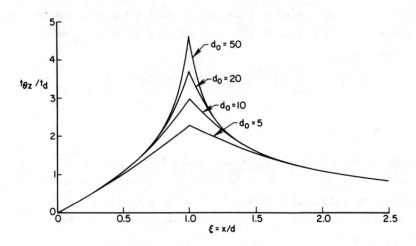

Figure 3 - Non-Dimensional Shear Stress for a Uniform Distribution of Screw Dislocations

For $d_0 = e \simeq 2.73$ this acquires its maximum

$$t_y^d/t_y^c \simeq 0.46 , \tag{3.19}$$

which indicates nearly 50 percent reduction in strength when there exists uniformly distributed dislocations over such a small length as d = 1.06a. In the case d = 4.5 x 10^{-4} cm for Fe.crystal, (3.18) gives

$$t_y^d/t_y^c = 2.94 \, x \, 10^{-4} , \tag{3.20}$$

as compared to the observed value [13] 2.2 x 10^{-4}. d = 4.5 x 10^{-4} cm used is one half of the length over which dislocations were distributed in inverse pile-up, as observed in the experiments of Ohr and Chang [12].

Clearly a better estimate of these values and N_{max} require finding n(x) for which $t_{\theta z}$ = const. in (3.12). The solution of this integral equation is not available at present.

4.   ANTIPLANE CRACK

The classical elasticity solution for a line crack in a plate under anti-plane shear loading at infinity is well known [11].

$$\sigma = \sigma_{23}-i\sigma_{13} = \sigma_0\bar{z}(\bar{z}^2-c^2)^{-1/2} , \qquad (4.1)$$

where σ_0 is the applied shear, $2c$ is the crack length and $z = x_1+ix_2$, $\bar{z} = x_1-ix_2$ are the rectangular coordinates in the $x_3 = $ const.-plane, Figure 4.

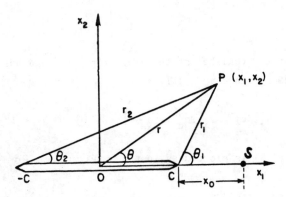

Figure 4 - Crack Subject to Anti-Plane Shear (Mode III)

Employing (4.1) in (3.3), we determine the nonlocal stress field

$$t = (\pi\varepsilon/2r_1)^{1/2}e^{-r_1/\varepsilon}(C_1e^{i\theta_1/2}+C_2e^{-i\theta_1/2}) + \sigma , \qquad (4.2)$$

where (r_1,θ_1) are the plane polar coordinates with the origin at the right-hand crack tip. The boundary condition at the crack tip surface requires that

$$\lim_{r\to0} t_{rz} = 0 , \qquad (4.3)$$

where t_{rz} is calculated by using (3.5) and (4.2). This gives $C_2 = 0$ and $C_1 = -(c/\pi\varepsilon)^{1/2}\sigma_0$. Consequently, [6],

$$t_{z\theta}-it_{zr} = \sigma_0(c/2r_1)^{1/2}[(2r^2/cr_2)^{1/2}e^{i(-\theta+(\theta_2/2))}-e^{-r_1/\varepsilon}]e^{-i\theta_1/2}, \qquad (4.4)$$

where (r_2,θ_2) are the plane polar coordinates with the origin in the left-hand crack tip and (r, θ) are the plane polar coordinates with the origin at $x_1 = x_2 = x_3 = 0$.

Along the x_1-axis $(\theta = \theta_1 = \theta_2 = 0)$, $t_{z\theta}$ acquires its maximum near the crack tip. From (4.4), we have

$$T_\theta(\rho) = \sqrt{\gamma} \, t_{z\theta}/\sigma_0 = (\pi\varepsilon)^{1/2} t_{z\theta}/K_{III}$$

$$= (2\rho)^{-1/2} \left[(1+\gamma\rho)(1 + \frac{\gamma\rho}{2})^{-1/2} - e^{-\rho} \right] , \qquad (4.5)$$

where

$$\rho = r_1/\varepsilon , \qquad \gamma = \varepsilon/c , \qquad K_{III} = \sqrt{\pi c} \, \sigma_0 . \qquad (4.6)$$

It is clear that $t_{z\theta}$ vanishes at the crack tip $\rho = 0$ and has a maximum at $\rho = \rho_c = 1.2565$, since $\gamma \ll 1$ $(\gamma \leq 10^{-4})$.

$$t_{z\theta} = \sigma_0 (\varepsilon/c)^{-1/2} \left(\frac{1}{\sqrt{2\rho_c}} + \frac{1}{\sqrt{2\rho_c}} \right)^{-1} \qquad (4.7)$$

$T_\theta(\rho)$ given by (4.5) is plotted against ρ in Figure 5. From this it is clear that:

(i) The stress field vanishes at the crack tip. This is in contradiction to the classical result that the stress is infinite at the tip.

(ii) Fracture begins when $t_{z\theta max} = t_y$, where t_y is the yield stress. From (4.7) we calculate

$$K_c/t_y = 3.9278 \sqrt{e_0 a} , \qquad (4.8)$$

where $K_c = \sqrt{\mu c} \, \sigma_{0c}$ is the critical fracture toughness. Using $e_0 = 0.39$, we calculate K_c-values for few materials, and compare with $K_c = (4\mu\gamma_s)^{1/2}$ used in atomic theories, where γ_s is the surface energy.

Table 1 - Stress Intensity Factors, K_c/t_y $(10^{-3} cm^{1/2})$

Material	Classical	Present	Experiment
A (fcc)	1.11	0.49	0.3
Cu (fcc)	3.86	0.47	0.66
Fe (bcc)	1.04	0.42	0.23

From these it is clear that the present estimates are reasonably close to the observations of Ohr and Chang [12] even though further considerations are necessary to allow for the presence of dislocations in the vicinity of the crack tip.

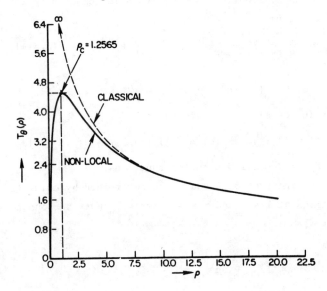

Figure 5 - Non-Dimensional Shear in Screw Dislocation

(iii) The maximum stress occurs at $\rho = \rho_c$. This implies that the fracture begins at a point *ahead* of the crack tip (not at the tip). This is against our previous understanding that fracture begins at the tip. The distance ρ_c is very small but not zero. With the dislocation accumulation, it may be modified, becoming larger than the present value of ρ_c.

5. EDGE DISLOCATION

Classical stress field for the edge dislocation may be expressed in the form

$$\underset{\sim}{\Theta}_\sigma = \frac{i\mu b}{2\pi(1-\nu)r} (e^{i\theta}-e^{i\theta}) \ , \qquad \underset{\sim}{\Phi}_\sigma = \frac{i\mu b}{2\pi(1-\nu)r} (e^{-i\theta}+e^{-3i\theta}) \ , (5.1)$$

where ν is the Poisson's ratio and

$$\underset{\sim}{\Theta}_\sigma = \sigma_{11} + \sigma_{22} \ , \qquad \underset{\sim}{\Phi}_\sigma = \sigma_{22} - \sigma_{11} + 2i\sigma_{12} \ . \qquad (5.2)$$

Differential equations (2.8) are equivalent to

$$(1-\varepsilon^2\nabla^2)\Theta_t = \Theta_\sigma \quad , \qquad (1-\varepsilon^2\nabla^2)\Phi_t = \Phi_\sigma \quad . \tag{5.3}$$

Integrations of (5.3) requires finding the solution of a differential equations of the form

$$\frac{\partial^2 g_n}{\partial\rho^2} + \frac{1}{\rho}\frac{\partial g_n}{\partial\rho} + \frac{1}{\rho^2}\frac{\partial^2 g_n}{\partial\theta^2} - g_n = -\frac{1}{\rho}e^{in\theta} \quad , \qquad (n = \pm 1, -3) \quad , \tag{5.4}$$

writing

$$g_n(\rho,\theta) = f_n(\rho)\exp(in\theta) \quad , \qquad \rho = r/\varepsilon \quad , \tag{5.5}$$

the general solution of (5.4) is found to be

$$f_n(\rho) = AI_n(\rho) + BK_n(\rho) + \int^{\rho} [I_n(z)K_n(\rho) - I_n(\rho)K_n(z)]dz \quad , \tag{5.6}$$

where I_n and K_n are modified Bessel's functions.

Constants of integration A and B are determined by using the regularity conditions at $r = 0$ and $r = \infty$, namely f_n must be bounded for $\rho = 0$ and $\rho = \infty$,

$$f_n = \int_0^{\rho} I_n(z)K_n(\rho)dz + \int_\rho^{\infty} I_n(\rho)K_n(z)dz \quad . \tag{5.7}$$

For $n = \pm 1$, these integrations can be carried out so that

$$f_{\pm 1} = \frac{1}{\rho}[1-\rho K_1(\rho)] \quad , \qquad f_{\pm 3} = K_3(\rho)\int_0^{\rho}I_3(z)dz + I_3(\rho)\int_\rho^{\infty}K_3(z)dz \quad . \tag{5.8}$$

Consequently,

$$\Theta_t = \frac{i\mu b}{2\pi(1-\nu)\varepsilon}\frac{1}{\rho}[1-\rho K_1(\rho)](e^{i\theta}-e^{-i\theta}) \quad ,$$

$$\Phi_t = \frac{i\mu b}{2\pi(1-\nu)\varepsilon}\left\{[\frac{1}{\rho} - K_1(\rho)]e^{-i\theta} + f_3(\rho)e^{-3i\theta}\right\} \quad . \tag{5.9}$$

In polar coordinates, we have

$$t_{rr}+t_{\theta\theta} = \Theta_t \quad , \qquad t_{\theta\theta} - t_{rr} + 2it_{r\theta} = \Phi_t e^{2i\theta} \quad , \tag{5.10}$$

from which we determine the stress field.

$$t_{rr} = - \frac{\mu b}{4\pi(1-\nu)\epsilon} \left[\frac{1}{\rho} - K_1(\rho) + f_3(\rho) \right] \sin\theta \ ,$$

$$t_{\theta\theta} = - \frac{\mu b}{4\pi(1-\nu)\epsilon} \left[\frac{3}{\rho} - 3K_1(\rho) - f_3(\rho) \right] \sin\theta \ , \qquad (5.11)$$

$$t_{r\theta} = \frac{\mu b}{4\pi(1-\nu)\epsilon} \left[\frac{1}{\rho} - K_1(\rho) + f_3(\rho) \right] \cos\theta \ .$$

Clearly, the stress field vanishes at $\rho = 0$ and $\rho = \infty$. Moreover, along the x-axis, $t_{rr} = t_{\theta\theta} = 0$ and $t_{r\theta}$ acquires its maximum at $\rho = \rho_c \simeq 1.2$.

In Figure 6, we display graphically, $T_E(\rho) = 4\pi(1-\nu)\epsilon t_{r\theta}/\mu b$ at $\theta = 0$, as a function of ρ. We have

$$T_{Emax} \simeq 0.3987 \ , \qquad \rho_c \simeq 1.2 \ . \qquad (5.12)$$

Consequently, fracture occurs when

$$t_y^c/\mu = 0.3987 \ \frac{b}{4\pi(1-\nu)\epsilon} \ . \qquad (5.13)$$

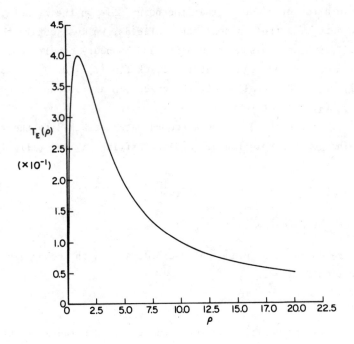

Figure 6 - Non-Dimensional Shear Stress in Edge Dislocation

If we compare this with (3.10), we see that this result becomes identical to (3.10) if we replace $2(1-\nu)$ by 1. Consequently, the theoretical strength of a pure crystal subject to an in-plane shear will differ, at most, by a factor of $\frac{1}{2}(1-\nu)^{-1}$ from that when it is subject to an anti-plane shear.

6. INTERACTION OF A SCREW DISLOCATION WITH A CRACK

Suppose that a homogeneous and isotropic elastic solid of infinite extent contains a crack at $|x_1| \leq c$, $x_2 = 0$, $|x_3| < \infty$ where x_k are rectangular coordinates. A screw dislocation which lies parallel to the x_3-axis intersects the $x_3 = 0$-plane at a point $S(x_1 = \xi, x_2 = \eta)$. The solid is subject to a constant anti-plane shear at $x_2 = \pm\infty$. The classical solution of this problem was given by Louat [14]. We quote here our own solutions [6][1] which is more suitable for the present treatment

$$\sigma \equiv \sigma_{23} - i\sigma_{13} = \frac{1}{\sqrt{z^2-c^2}}\left[\sigma_0\bar{z} + \frac{\mu b}{2\pi}\left(1 + \frac{\sqrt{\xi^2-c^2}}{\bar{z}-\xi}\right)\right], \qquad (6.1)$$

where we also set $\eta = 0$ so that the point S is on the x_1-axis, Figure 4.

To determine the nonlocal stress fields, we must obtain the solution of (3.3), subject to some regularity and boundary conditions,

(i) u_z must be regular at the crack tip ($r_1 = 0$, $\theta_1 = 0$);

(ii) $t_{rz} = 0$ for all θ, at the crack tip and at the dislocation;

(iii) $t_{23} = \sigma_0$ at $x_2 = \pm\infty$.

Here (r_1, θ_1) are the plane polar coordinates attached to the right crack tip. The general solution of (3.3) satisfying (i) and (iii) is of the form

$$t = t_{23} - it_{13} = (\pi\varepsilon/2r_1)^{1/2}e^{-r_1/\varepsilon}(C_1e^{i\theta_1/2} + C_2e^{-i\theta_1/2}) +$$

$$K_1(r_d/\varepsilon)(C_3e^{i\theta_3} + C_4e^{-i\theta_3}) + \sigma, \qquad (6.2)$$

where (r_3, θ_3) are the plane polar coordinates with the origin at the dislocation S, i.e.,

$$r_d e^{i\theta_3} = r_1 e^{i\theta_1} - x_0, \qquad (6.3)$$

and $K_1(z)$ is the modified Bessel's function. Constants of integrations C_α are determined by using boundary conditions (ii). This leads to [6]:

---

[1]

σ_{23} given here is equal to $\sigma_{23} + \sigma_0 + \sigma_\alpha(x_1)$ of Reference [6] and σ_{13}'s are identical.

$$t = (\pi\varepsilon/2r_1)^{1/2} e^{-r_1/\varepsilon} C_1 e^{i\theta_1/2} - \frac{\mu b}{2\pi\varepsilon} K_1(r_d/\varepsilon) e^{i\theta_3 + \sigma} \quad , \tag{6.4}$$

where

$$C_1 = -(c/\pi\varepsilon)^{1/2} \left\{ \sigma_0 + \frac{\mu b}{2\pi c} \left[1 - \left(1 + \frac{2c}{x_0} \right)^{1/2} \right] \right\} \quad . \tag{6.5}$$

By means of (3.5), we obtain components of the stress field in polar coordinates (r,θ)

$$t_{\theta z} - it_{rz} = t^c + t^{dc} \quad , \tag{6.6}$$

where

$$t^c \sqrt{\alpha}/\sigma_0 = (2\rho_1)^{-1/2} \left[\rho(\rho_2/2\alpha)^{-1/2} e^{-i\theta + i(\theta_2 - \theta_1)/2} - e^{-\rho_1 - i\theta_1/2} \right] \quad , \tag{6.7}$$

$$t^{dc} \sqrt{\alpha}/\sigma_0 = (2\rho_1)^{-1/2} \left\{ (\alpha\rho_2/2)^{-1/2} e^{i(\theta_2 - \theta_1)} - \left[1 - \left(1 + \frac{2}{\alpha \bar{x}_0} \right)^{1/2} \right] e^{-\rho_1 - i\theta_1/2} - \right.$$

$$\left. (2\rho_1/\alpha)^{1/2} K_1(\rho_d) e^{i(\theta_3 - \theta_1)} + (\alpha\rho_2/2)^{-1/2} \left(1 + \frac{2}{\alpha\bar{x}_0} \right)^{1/2} \left(\frac{\rho_1}{\bar{x}_0} e^{-i\theta_1} - 1 \right) e^{i(\theta_2 - \theta_1)} \right\} \quad , \tag{6.8}$$

with definitions

$$t^c = t^c_{\theta z} - it^c_{rz} \quad , \qquad t^{dc} = t^{dc}_{\theta z} - it^{dc}_{\theta z} \quad ,$$

$$\bar{x}_0 = x_0/\varepsilon \quad , \qquad \beta = \mu b/2\pi c\sigma_0 \quad , \qquad \alpha = \varepsilon/c \quad , \tag{6.9}$$

$$\rho_1 = r_1/\varepsilon \quad , \qquad \rho_2 = r_2/\varepsilon \quad , \qquad \rho_d = r_d/\varepsilon \quad .$$

Plane polar coordinates $(r_\alpha, \theta_\alpha)$ are shown in Figure 4.

Along the crack line $\theta_1 = \theta_2 = \theta = 0$ and $\theta_3 = 0, \pi$ so that $t_{rz} = 0$ and $t_{\theta z}$ is given by

$$t_{\theta z} = \sigma_0 (T_1 + \beta T_2)/\sqrt{\alpha} \quad , \tag{6.10}$$

$$T_1 = (2\rho_1)^{-1/2} \left[(1 + \alpha\rho_1) \left(1 + \frac{\alpha\rho_1}{2} \right)^{-1/2} - e^{-\rho_1} \right] \quad , \tag{6.11}$$

$$T_2 = (2\rho_1)^{-1/2} \left\{ \left(1 + \frac{\alpha\rho_1}{2} \right)^{-1/2} - \left[1 - \left(1 + \frac{2}{\alpha\bar{x}_0} \right)^{1/2} \right] e^{-\rho_1} - \right.$$

$$\left. \text{sgn}(\rho_1 - \bar{x}_0)(2\rho_1/\alpha)^{1/2} K_1(|\rho_1 - \bar{x}_0|) + \left(1 + \frac{\alpha\rho_1}{2} \right)^{-1/2} \left(1 + \frac{2}{\alpha\bar{x}_0} \right)^{1/2} \left(\frac{\rho_1}{\bar{x}_0} - 1 \right)^{-1} \right\} \quad . \tag{6.12}$$

When the dislocation is absent, we have $t^{dc} = 0$ so that t^{dc} is the shear stress arising from the interaction of the dislocation with the crack. At the crack tip, $\rho_1 = 0$ and we have

$$t_{\theta z}/\sigma_0 \beta = \frac{1}{\alpha} K_1(\bar{x}_0) \quad . \tag{6.13}$$

This shows that for positive dislocation (b > 0), the shear stress is
positive and therefore the crack tip will tend to close up for b > 0.
For b < 0, the opposite will occur. However, the stress given by (6.13)
is very small for large x_0 and it becomes large when the dislocation is
very close to the crack tip.

The total non-dimensional shear stress $T = t_{\theta z}/\sigma_0$ versus ρ_1 is
plotted in Figure 7 for some values of $\bar{x}_0$ and $\alpha = 10^{-8}$, $\beta = 10^{-4}$. The
Burger's vector b is taken to be negative. The ratio of the maximum
stresses with and without a dislocation give

$$K_{ctot}/K_c \simeq 0.71 \ . \tag{6.14}$$

This implies that a dislocation located at a distance approximately one
or two lattice parameters away from the crack tip reduces the fracture
toughness by about 30 percent. Hence the theoretical K_c listed in
Table 1 will be reduced about 30 percent, bring the present estimates
even closer to the experimental values.

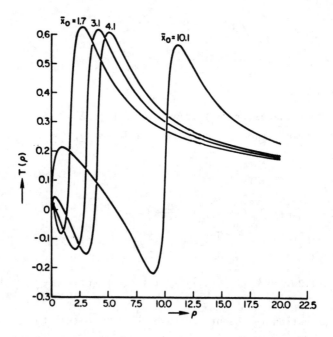

Figure 7 - Total Shear Stress (Crack and Dislocation)

A more realistic picture requires the consideration of the distribution of dislocation over a few microns or so distance, away from the crack tip. This picture is that observed in electron microscopy [12].

A precise calculation of the combined stress due to a crack and a distribution of dislocations requires the consideration of the dislocation pile-up. Mathematical solution of this problem is rather involved. However, based on the localization of the stress in plate due to a crack alone and the observation of the locations of the dislocations in electron microscopy, we can present an approximate calculation here. The present calculations show that the maximum shear stress due to crack occurs very close to the crack tip (only one or two lattice parameters away from the tip). Ohr and Chang's [12] observation in an electron microscope indicated a dislocation-free zone of approximately 1 to 5 μm near the crack tip. At this distance, the shear stress due to crack alone is very nearly equal to the applied shear stress σ_0. Therefore, we can superpose the stress distribution due to dislocation pile-up with the applied stress σ_0.

For a uniform distribution of dislocations, the total maximum stress is

$$t_{tot} = \sigma_0 + \frac{\mu bN}{2\pi\epsilon} \frac{\ell n d_0}{d_0} , \qquad (b > 0) . \qquad (6.15)$$

If we set $t_{tot} = t_y$ and write $K_g = \sqrt{\pi c}\, \sigma_0$ and $K_c = \sqrt{\pi c}\, t_y$, (6.15) may be put into the form

$$K_g/K_c = 1 - \kappa N \frac{\ell n d_0}{d_0} , \qquad (6.16)$$

where

$$\kappa = \mu b/2\pi\epsilon\, t_y , \qquad (b > 0) , \qquad (6.17)$$

is a *material constant*. From (6.16), it is clear that for $b > 0$, the fracture toughness K_c decreases to K_g in proportion to the total number of dislocations. For iron, we calculate $\kappa = 1.70$. Experimentally observed length of the dislocation zone for Fe. crystal was in the neighbourhood of 9 μm [12]. However, the distribution of dislocations was not uniform. Using $N = 1514$ obtained in Section 3, by means of equation (6.3) we calculate $K_g/K_c \simeq 0.32$. This result brings the calculated value 0.42 closer to the observed value 0.23 (see Table 1).

More realistic calculations require the determination of $n(x)$ from the integral equation for the dislocation pile-up leading to a constant yield stress distribution in an interval of length $2d$. In fact, such calculations are required for a $t_{\theta z}$ = const. distribution in the *yield region* around the crack tip rather than in an interval $|x| < d$, $y = 0$.

ACKNOWLEDGEMENT

   The present work was supported by the Office of Naval Research. The author is indebted to Dr. N. Basdekas for his encouragement and support. I wish to thank Victor Chen and A. Suresh for the computer work.

REFERENCES

1. ERINGEN, A.C., "Screw Dislocation in Nonlocal Elasticity", *J. Phys. D. Appl. Phys.*, Vol. 10, 1977, pp. 671-678.
2. ERINGEN, A.C., SPEZIALE C. and KIM, B.S., "Crack-Tip Problem in Nonlocal Elasticity", *J. Mech. Phys. Solids*, Vol. 25, 1977, pp. 339-35
3. ERINGEN, A.C., "Line Crack Subject to Shear", *Int. J. of Fracture*, Vol. 14, 1978, pp. 367-379.
4. ERINGEN, A.C. and SURESH, A., "Nonlocal Effects of Crack Curving", *Proc. of the Second Int. Sym. and Seventh Canadian Fracture Conf. on Defects, Fracture and Fatigue*, edited by G.C. Sih and J.W. Provan, Martinus Nijhoff Publishers, 1983, pp. 233-241.
5. ERINGEN, A.C., "On Differential Equations of Nonlocal Elasticity and Solutions of Screw Dislocation and Surface Waves", *J. Appl. Phys.*, Vol. 54, 1983, pp. 4703-4710.
6. ERINGEN, A.C., "Interaction of a Dislocation with a Crack", *J. Appl. Phys.*, Vol. 55, 1984, (to appear).
7. ERINGEN, A.C., "Linear Theory of Nonlocal Elasticity and Dispersion of Plane Waves", *Int. J. Eng. Science*, Vol. 10, 1972, pp. 425-435.
8. ERINGEN, A.C., "Nonlocal Polar Field Theories", *Continuum Physics*, Vol. IV, Academic Press, New York, 1976, pp. 204-267.
9. HIRTH, J.C. and LOTHE, J., *Theory of Dislocations*, McGraw-Hill, New York, 1968, p. 60.
10. KELLY, A., *Strong Solids*, Oxford, 1956, p. 12.
11. SNEDDON, I.N. and LOWENGRUB, M., *Crack Problems in the Classical Theory of Elasticity*, John Wiley, New York, 1969, p. 38.
12. OHR, S.M. and CHANG, S.J., "Direct Observations of Crack Tip Dislocation Behavior During Tensile and Cyclic Deformation", *Technical Report*, Oak Ridge National Laboratory.
13. JAYATILKA, A.S., *Fracture of Engineering Brittle Materials*, Applied Science Publishers, London, 1979, p. 46.
14. LOUAT, N.P., "The Interactions of Cracks and Dislocations in Screw Dislocations", *Proc. Int. Conf. on Fracture*, Sendai, Japan, 1965, pp. 117-132.

THE MICROMECHANICS OF FATIGUE CRACK INITIATION

J.W. PROVAN
Department of Mechanical Engineering
McGill University
Montreal, Quebec, Canada

1. INTRODUCTION

Over the last few years there has been an increasing interest in assessing the reliability of involved and expensive engineering structures and components being subjected to fatigue loading situations. The approach being taken in these studies has been and still is to describe, in a statistical manner, the microstructural material degradation processes taking place in polycrystalline materials and to infer from analytic, numeric and laboratory studies the reliability of the components and structures that utilize the materials in question prior to them leaving the design stage. The motivation behind these studies is to:

(a) be in a position to supply to a prospective customer, definitive instructions as to safety, inspection intervals, spare part availability, life cycle costs, etc., and

(b) be confident in these estimates based on accelerated and small scale laboratory testing and numerical algorithms.

This has been the expressed aim of an ongoing research program being currently undertaken in the Micromechanics of Solids Laboratory of McGill University. To date, this program has met with varied success but still requires further study prior to it being implemented with confidence on a large scale. One of the most promising outcomes of this research endeavour has been the Provan reliability law [1] whose derivation was based entirely on modelling the fatigue crack initiation and propagation processes by a statistical interference technique and by a Markov linear birth stochastic process, respectively. This reliability law in its failure density form is expressed as:

$$p_{N_p}(i) = \frac{\mu_{a_0}}{\sqrt{2\pi}V_{ac}} \exp\left\{ \lambda i - \frac{(\mu_{a_0}\exp[\lambda i]-\mu_{af})^2}{2V_{ac}} \right\} , \qquad (1)$$

with its associated cumulative expression being:

$$P_{N_p}(j) = \int_{-\infty}^{j} p_{N_p}(i)\,di = \frac{1}{2}\left\{ \mathrm{erf}\left(\frac{\mu_{a_0} e^{\lambda i} - \mu_{a_f}}{\sqrt{2V_{ac}}}\right) + \mathrm{erf}\left(\frac{\mu_{a_f}}{\sqrt{2V_{ac}}}\right)\right\}, \qquad (2)$$

where:

$$\mathrm{erf}(x) = \frac{2}{\sqrt{\pi}} \int_{0}^{x} e^{-z^2}\,dz \;,$$

$\quad V_{ac} \quad$ = variance of crack penetration,

$\quad \mu_{a_0} \quad$ = mean of crack initiation depth,

$\quad \lambda \quad$ = growth rate transition intensity and   (3)

$\quad \mu_{a_f} \quad$ = mean of critical crack depth.

Strictly speaking the reliability law is simply $R_j = 1 - P_{N_p}(j)$, but in keeping with common usage the expressions (1) and (2) are termed the reliability functions. The skewed to the left characteristic shape of the cumulative expression, for example, is illustrated by the solid line in Figure 1. Subsequent to the derivation of this reliability law, an extensive experimental program was entered into in order to check on the validity of these expressions [2,3]. They were performed on oxygen free high conductivity (O.F.H.C.) copper within the guidelines specified by the ASTM standard E466 [4], under strain control at a total strain amplitude of $\Delta\epsilon/2 = 0.003$ and at a frequency of 0.5 Hz. The data collected in an increasing life cycle format are indicated by the dots in Figure 1. By curve fitting the Provan reliability law to the experimental points, the empirical value of the fundamental parameter known as the transition intensity was deduced as:

$$\lambda_{emp} = 0.20 \times 10^{-3}/\text{cycle}\,; \qquad \frac{\Delta\epsilon}{2} = 0.003\,. \qquad (4)$$

As is observed, the agreement between the empirical and test results is reasonably good thereby confirming the validity of the approach being taken by this research investigation. However, as noted above, the whole idea of this research program is, that on the basis of a few accelerated laboratory tests and numerical algorithms, to be in a position *to infer* the reliability of both the material and, hence, the components manufacture from this metal. Two methods have previously been suggested [5,6]; one theoretical and the other experimental. The theoretical estimation procedure was shown in both [2] and [5] to seriously underestimate the number of cycles involved in initiating a fatigue crack and, as a result, the

experimental fractographic procedure also described in [5], is preferred
at present. Following this procedure, the experimental value of the
transition intensity was evaluated as:

$$\lambda_{exp} = 1.89 \times 10^{-3}/\text{cycle} ; \quad \frac{\Delta\varepsilon}{2} = 0.003 , \tag{5}$$

thereby giving a ratio $\lambda_{exp} \approx 9.5 \, \lambda_{emp}$.

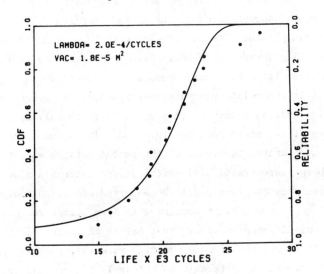

*Figure 1 - The Empirical Fit of the Provan Reliability Law to Fatigue
Data Points*

The reason for this order-of-magnitude discrepancy is now well understood and indeed has motivated the present investigation whose status is
reported in the subsequent sections of this presentation. The reason is
that *while the reliability expressions and the fractographic procedure
deal with the number of fatigue cycles required to propagate a crack, the
test results, illustrated in Figure 1, are for the combined initiation and
propagation fatigue life of the specimens.* The order-of-magnitude difference in the case of O.F.H.C. copper being tested in usual laboratory
conditions, therefore suggests that approximately 85-90 percent of the
total number of cycles required to fail the specimens are, in fact,
involved in *initiating* their respective fatigue cracks. This paper
therefore discusses the micromechanics of fatigue crack initiation along
with experimental procedures that may in the future enable us to detect,
in a nonarbitrary manner, the cycles involved in initiating a crack. When

this is accomplished, the Provan reliability law will be one step closer
to being implementable.

2. THE MICROMECHANICS OF FATIGUE CRACK INITIATION

2.1 Introduction

This section reviews the micromechanics and the theories of what is
known concerning the nature of fatigue crack initiation in polycrystalline
metals. A thorough understanding of the micromechanisms taking place in
a fatigued material is the first prerequisite to being able to model and,
hence, to eventually detect crack initiation. A definitive review of
the mechanisms associated with this phenomenon, especially in the case
of single crystals and relatively pure polycrystals, has recently been
presented in [7] and, as a result, this present review only concerns itself
with highlights of the microprocesses involved. It is understandable,
but somewhat unfortunate, that most of the recent fatigue mechanisms
research has been concentrated on lightly alloyed materials since the
fundamental processes can more easily be observed and discussed. The
same processes must, however, be occurring in more structured metals,
albeit on various different scales. What may be an acceptable crack initiation model for a relatively pure, ductile material, such as O.F.H.C.
copper, is not directly transferable to a heavily alloyed and brittle
material, such as gray cast iron, without augmenting the model with a
statement concerning the influence of precipitates and secondary phase
particles [8]. Unfortunately, much less is known concerning their influence on crack initiation than the corresponding influence of dislocations
and their substructures. As a result, the following review mainly applies
to relatively pure metallic polycrystals.

2.2 Low Cycle Strain Deformation

In the case of annealed single crystals that are being cyclically
strained so as to experience small amounts of plastic deformation per cycle
this plastic strain is almost exclusively the result of primary dislocations whose motion starts as soon as the first plastic strain cycle is
experienced. As the crystal is cycled, this dislocation associated damage
causes the monocrystals to at first rapidly harden and then to reach a
saturation stage or plateau in a manner illustrated in [9]. In [10], it
is made clear that the initial rapid hardening occurs by the mutual trapping

of edge dislocations and their formation into cellular bundles of dense
loop patches between which screw dislocations find it relatively easy
to glide. As the number of cycles increases these dislocation cells,
which are illustrated in Figure 2 and are homogeneously distributed throughout the monocrystals, gradually grow in number until a balance is reached
between the edge loop patches and the intervening matrix channels that
are transversed by the screw dislocations; a balance that is represented
by saturation. In certain materials, the onset of saturation is associated
with the dislocation cells forming into persistent slip bands (PSBs) whose
thin lamellae are parallel to the primary slip plane and cause deformation
to become inhomogeneous and strongly localized in the PSB-matrix interface.
The dislocation structure of the PSBs is significantly different from that
of the dislocation cells in that they essentially consist of cleared
channels interrupted by dense thin dislocation "hedges", approximately
1 - 1.5 µm apart at room temperature. The actual mechanism by which they
are formed is still an open question. Despite this, it is now accepted
that the phenomenon of PSB formation has fundamental consequences as far
as fatigue crack initiation is concerned and therefore they warrant a
separate discussion.

Figure 2 - Dislocation Cells in α-Iron (Reference [14])

2.3 Persistent Slip Bands (PSBs)

The formation of PSBs appears to be a general feature of face-centred cubic (fcc) relatively pure polycrystalline metals, such as copper, nicke and silver, and to a lesser extent, of body-centred-cubic (bcc) metals, such as α-iron, 0.1 wt% carbon structural steel and niobium [9,11-15] and hexagonal-closed-packed (hcp) metals that are being subjected to fatigue loading situations. For high cycle low amplitude fatigue, crack initiati has been generally accepted as occurring in a transgranular manner by slip-band cracking at the PSB-matrix interface and intergranularly (non-PSB related) for low cycle high amplitude fatigue [16]. Quantitative evidence of the increasing plastic strain amplitude is given in [17]. Alternatively, other work has shown that crack initiation at grain boundaries, or for that matter at any other interface such as an inclusion-matrix interface or a twin boundary, may also occur in high cycle fatigue by a mechanism of grain boundary impingement by PSBs [13]. In single crystals [18], the wall or ladder structure of PSBs is a result of edge dislocations being deposited, as schematically shown in Figure 3(a), at the PSB-matrix interfaces. Under the combined action of the applied stress and of the repulsive internal stress due to these interfacial dislocations, the latter emerge, in the case of single crystals, a extrusions on both sides of the PSB, as indicated in Figures 3(b) and 4. Similar considerations apply to PSBs in polycrystals with respect to their impingement with grain boundaries, for example. Due to the rigidity of the surrounding material, grain boundaries only allow constrained displacements either in the nature of localized shearing of the PSB walls, illustrated by Figure 5(a), or in the pile-up of the PSB-matrix interface dislocations at the grain boundaries, schematically shown in Figure 5(b).

In the case of low cycle high amplitude fatigue, the dislocation cell structures are generally throughout a specimen of polycrystalline metal and the localized inhomogeneous deformation associated with PSBs is no longer observed. Instead, microcracks tend to form intergranularly at high-angle grain boundaries as a result of small steps being formed in tension which are not completely cancelled by the compressive side of the reversed cycling [19]. This step gradually increases until a very sharp notch root developes and acts as a stress raiser from which the crack grows along the grain boundary into the material in question. Depending on a variety of factors including environmental, either mechanism, i.e.,

the transgranular, as illustrated in Figure 6 in the case of 99.93 wt%
iron [17], or intergranular, can eventually lead to the nucleation of
localized fatigue microcracks. Upon the coalescence or growth of these
microcracks into a crack which:

   (i) changes the direction of the predominant stress from that of the
localized shear stress to that of the local applied normal stress, and;

   (ii) causes the stress field in the neighbourhood of this crack to
increase to such an extent that other cracks in its vicinity become, so
to speak, dormant,

then this may be taken as a realistic mechanism and micromechanics definition of fatigue crack initiation. Notice that this relatively well accepted
model is not restricted either to the surface of the specimen or component
and is indeed applicable to a microstructural description of what occurs in
advance of a propagating crack. A further point is that it highlights the
statistical and stochastic nature of fatigue crack initiation that, in
general, has been central to the micromechanics approach to fatigue failure
and does not depend on a physical dimension for its definition. Finally,
it is consistent with both Stage 1 cracking as defined in [20] and the
interpretation of initiation as being associated with microcrack
coalescence [21].

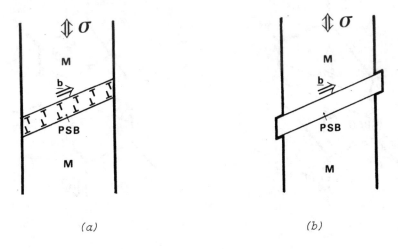

(a)   (b)

Figure 3 - PSB Formation of Extrusions (Reference [13])

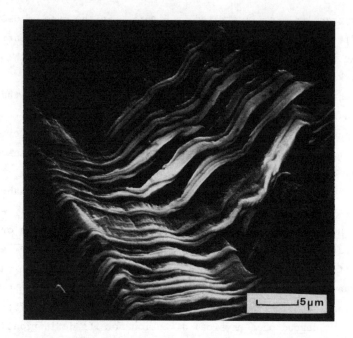

Figure 4 - Surface PSBs in α-Iron (Reference [14])

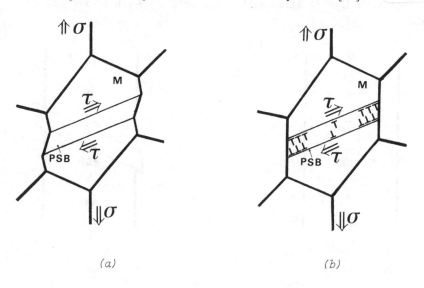

(a)   (b)

Figure 5 - Interaction of a PSB with Grain Boundaries (Reference [13])

Figure 6 - Transgranular Microcracks in High Purity Iron (Reference [17])

A tremendous amount of work has already been carried out in order to gain an understanding of the micromechanics of fatigue crack initiation. However, a large amount is still to be carried out especially as it relates to metals of engineering importance whose microalloying and heat treatments practically define their applicability in fatigue load carrying situations [22-24].

3.   TECHNIQUES FOR DETECTING CRACK INITIATION

3.1 Introduction

No one technique has emerged over the last decade as being the one and only for the monitoring of fatigue crack initiation. If there was, then a lot of the uncertainty concerning fatigue life estimates and reliability would have been resolved. As it is, there is a great deal of current interest in nondestructive-evaluation (NDE) methods in the hopes that they will not only help in strengthening the utilization of damage tolerant design methods but will also apply to the nonarbitrary determination of fatigue crack initiation. It is this latter context that is of most interest in this section.

There have been a number of reviews on this subject with those given in [25,26] being appropriate to the present discussion. Table 1 lists the better known methods and their range of applicability as discussed in [25] and hence this section will concentrate on some other methods that have been suggested in the recent literature as being possible methods of detecting initiation. Their choice, however, is not entirely objective since they represent some of the past and present experimental interests of the author. For convenience, the methods that are briefly discussed in this section are listed along with some pertinent comments in Table 2.

Table 1 - Applicability of Various NDE Techniques to Monitoring Dislocations and Microcrack Formation and Growth During Fatigue (Reference [25])

Technique	Dislocation Density/ Arrangement	Microcrack Formation/ Growth
Acoustic attenuation by dislocations	Prob. dem.	-
Acoustic transmission/reflection by interfaces	-	Dem.
Acoustic scattering	-	Prob. Dem.
Acoustic harmonic generation	Dem.	Dem.
Acoustig emission	Prob. dem.	Prob. Dem
Surface topography (acoustical)	Poss.	Poss.
Surface topography (optical)	Dem.	Dem.
Photostimulated exo-electron emission	Dem.	Dem.
Positron annihilation	Dem.	-
Eddy-current techniques	Poss.	Prob. Dem.
Gage concept	Poss.	Dem.

Table 2 - Experimental Techniques and Their Nonarbitrary Crack Initiation Detection Potential

CLASS	METHOD	COMMENTS	RATING /10
Optical	Light Microscopy	System resolution defined crack initiation size.	2
	Photoelasticity, Photoplasticity	Relies on empiricism and specific material properties.	4-5
	White Light Interferometry	Has a good resolution and potential.	8
	Holographic Interferometry	Excellent accuracy, poor magnification.	3
Electrical Resistive	Strain Gages	Good accuracy and easy to use.	4
	Potential Drop	Have noise problems.	6
Electron Beam	In-situ SEM	In vacuum. Excellent potential.	9
	Electron Channeling	In vacuum. Excellent potential.	9
	In-situ TEM	In vacuum. Excellent potential.	9
Mechanical	Compliance	The prospects are good.	7
	First Order Changes	Very low sensitivity.	0
Energy Absorbing	Radiography	Very low sensitivity.	0
	X-Ray Lang Camera	Has specimen restrictions and limited applicability.	1-2
	Ultrasonics	Industrially suited and improving.	7
Energy Emitting	Acoustic Emission	Has given somewhat disappointing results.	6
	Thermal Energy Release	The prospects are good.	5
	Exoelectron Emission	In air and vacuum. Excellent potential.	9

3.2 Optical Techniques

3.2.1 Light Microscopy. This method is one of the oldest and still productive techniques available for detecting a fatigue induced crack of an experimental system dependent size. Since it relies on both the quality of the test apparatus and the crack being on the surface of the specimen, it is not entirely suited to the detection of the micromechanics interpretation of crack initiation. It has, however, been successfully used to monitor, among other things, crack initiation when it has been interpreted as a crack of certain size [27].

3.2.2 Photoelastic Surface Coatings and Photoplasticity. This method concerns the measurement of elastic and plastic strains on the surface of specimens by means of the elastic strain in an optically active coating or varnish based on the measurement of birefringency. It does permit the use of the real materials with the sensitivity of the method being proportional to the optical properties and thickness of the transparent birefringent coating layer [28]. Unfortunately, since the accuracy of the photoplasticity method in the inelastic states of strain depends on a number of measured material constants and empirical relationships, the assessment of a particular system's sensitivity not only is crucial but also limits this technique to cracks larger than those usually of interest in crack initiation studies.

3.2.3 White Light Interferometry. Interferometry was successfully applied in [19] for the study of low cycle high amplitude crack nucleation and characterization of the locations where they form. This white light two beam interference procedure with a resolution of approximately 0.4 µm clearly showed the gradual formation, with increasing stress reversals, of a grain boundary step or very sharp notch root from which the crack grew in a manner previously described. Although this procedure has (to the author's knowledge) not been specifically applied to the detection of the above definition of initiation it most certainly has this potential since there is a possibility of observing a change in failure mechanism between stage I and stage II propagating crack.

3.2.4 Holographic Interferometry. The advent of laser technology has opened this new class of measurement technique which is still in rapid growth and offers promising possibilities for the future. The first applications of holography to fracture used both real time [29] and double exposure [30] interferometry, techniques that have occupied the talents of many researchers, especially in the NDE field [31], since then. The method which permits the measurement of general three-dimensional displacements and thus strains, on any visually accessible diffusely reflective surface, can detect the presence of surface or near surface cracks and their growth to a sensitivity of 0.25 μm, although their exact outline may not be dis-cernible. Recently, a holographic interferometric study of the J-integral and C.O.D. as they apply to elastic-plastic fracture mechanics was carried out [32]. A photograph of an interferogram of a compact tension specimen under an axial load of only 115 N is shown in Figure 7. It clearly shows the applicability of this method to the detection of extremely small dis-placements under extremely small loads, the quantities required for the possible detection of fatigue crack initiation. It has this potential but, however, the technical problems associated with the set up of this experimental apparatus rule the procedure out as one that can be routinely performed in a laboratory.

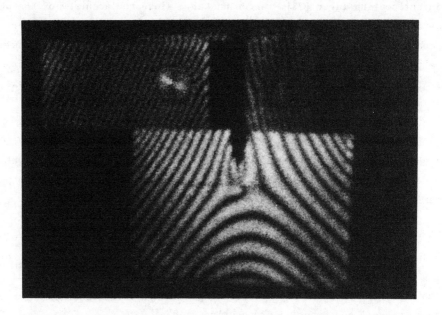

*Figure 7 - Photograph of the Interferogram of a CT-Specimen under a
          Load of 115N.*

3.3 Electrical Resistivity Procedures

3.3.1 Strain Gages. Electrical strain gages can either be applied
directly to test specimens or components to measure the surface strains
at various locations or can be bonded to flexible beams such as in the case
of the familiar clip-on gage. They have been recently used to measure the
back-face strains in a compliance method of evaluating crack length [33]
and the strain near the crack tip for the purpose of detecting ductile
crack growth [34]. In this latter study, crack initiation was identified
as the attainment of the maximum strain but since no elastic-plastic analysis is as yet available and since it is a surface measurement there is
reason to be cautious about this conclusion. Still, strain gage techniques
are improving at a steady rate and it is not beyond the realm of possibility
that they will prove most useful in detecting crack initiation, certainly
in the macroscopic, laboratory and component reliability senses.

3.3.2 Electrical Potential Difference Methods. Very popular laboratory techniques for the measurement of crack length are based on the phenomenon of potential drop during crack extension. There are three variations
of this method, namely the AC, the DC and the thin metallic conductor layer
bonded to the surface of conducting and nonconducting materials. All rely
upon the increase in electrical resistance of the conductor as its cross-
sectional area is reduced by crack growth and have a possible crack size
resolution of $\approx$ 10 μm. The major difficulties, as discussed in [35-37],
in the operation of these techniques arise from the nature of the crack
whose extension is to be measured; in particular, from the shape of the
crack front, the electrical contact across the crack faces (due sometimes
to oxidation layers) and the presence of the plastic zone. The major
advantages of the DC method are that it does not depend on advanced
electronics and, for certain specimen geometries such as the CT-specimen,
is a well known and established technique. On the other hand, however,
the relationship between the potential drop and crack length is rather
complex and the technique suffers from all the problems associated with
the handling of low level signals. The AC method is known for its ease
of calibration and signal amplification along with the lack of size dependence of the results. Problems associated with the AC method are those
of lead interaction and electronic stability. All three variations have
the potential to detect crack initiation in the case of very carefully
controlled laboratory experimentation.

3.4 Electron Beam Techniques

3.4.1 In-situ Scanning Electron Microscopy. Since fatigue initiation
is, for the most part, an intrinsically local surface phenomenon, direct
observation in a scanning electron microscope (SEM) has become a powerful
means of gaining an understanding of the micromechanisms taking place
during this phase of fatigue induced damage. It has been successfully
applied in [38] to measure the initial mode II crack growth mechanism
which, in the context being discussed in this paper, corresponds to the
second phase of stage I fatigue fracture. As a result, this technique
holds tremendous promise as far as small specimen, in vacuum crack initiation studies are concerned, as evidenced by its current success in determining the microstructural processes involved in the fatigue threshold
and crack propagation stages of material damage [39]. The literature
abounds with such studies and indeed whole symposia have been devoted
to these investigations.

3.4.2 Electron Channeling. The phenomenon of selective electron absorp-
tion by the crystalline structure of individual grains of a polycrystalline
metal is known as electron channeling [40]. In principle, both the initiation and propagation phases may be studied although, to date, studies
have tended to concentrate on crack propagation. Although most often
studied in conjunction with an SEM, other systems having suitable scanning
and detection capabilities may be used. Electron channeling may be applied
to the detection of crack initiation in either the channeling contrast or
channeling pattern modes of operation, respectively. As mentioned above and
elsewhere, cyclic deformation causes the regular structure of metals to
develop a microstructure observable by channeling contrast which measures
misorientations as small as $0.2°$ of dislocation cells relative to their
neighbours. This procedure has been successful in determining the effects
of cyclic stress intensity and environment on the plastic zone size and in
measuring energy dissipation during crack propagation. It is this latter
capability that may prove useful in nonarbitrarily distinguishing initiation from propagation. Selected area electron channel patterns, on the
other hand, are formed in a small surface area by rocking the electron
beam of the SEM about a focal point at the specimen surface. It is not
a straight forward procedure and, since a major limitation is its spatial
resolution, channeling contrast has been preferred when studying the

deformation near the crack tip. This technique has tremendous prospects as far as the detection of crack initiation is concerned.

3.4.3 In-situ Transmission Electron Microscopy (TEM). Apart from its more traditional role, namely that of replica fractography [41-44], TEM and high voltage electron microscopes (1MV), have more recently, however, been used to assess the dislocation structures and their motions under in-situ fatigue loading [45,46]. It is primarily by the use of these instruments that the dislocation cell and PSB substructures discussed in Section 3.3 have been ascertained. In a very decisive manner it has also been recently used to observe crack tip dislocation behaviour during cyclic deformation [47], a study that clearly showed the importance of local mode II crack propagation consistent with the previous statement that the proposed stage I initiation definition may also be applicable to a description of the microstructural processes taking place in front of a propagating crack.

3.5 Mechanical Procedures

3.5.1 Material Compliance Changes. In [17], a new technique was introduced for assessing microcrack initiation using a purely mechanical procedure. It is based upon the detection of small changes in the material compliance of a specimen at the point of the low cycle hysteresis loop where strain reversal from maximum tension takes place, i.e., it is a second order change in the mechanical properties. Defining E_T as the slope of the stress-strain curve during this portion of the unloading cycle, the results shown in Figure 8, where E_T is plotted against $\varepsilon_{pa} = 4 N (\Delta\varepsilon_p/2)$, the accumulated plastic strain, clearly indicates a decrease in the E_T parameter just prior to the smallest crack of 2 μm being discernible at the point indicated by A. While the findings of this investigation are still preliminary, the prospects of assessing crack initiation by a precise utilization of a standard closed-loop material testing facility without resorting to more involved companion crack initiation measuring techniques warrents further study of this procedure.

3.5.2 First Order Changes in Mechanical Properties. Changes in the shape of the stress-strain hysteresis loop similar to those used in [48] are considered by this author, among others, to be too insensitive to detect crack initiation as it is being interpreted in this presentation.

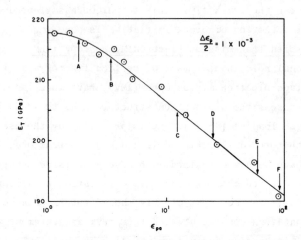

Figure 8 - Compliance Change as an Indication of Crack Initiation
(Reference [17]).

3.6 Energy Absorbing Procedures

3.6.1 Radiography. This method, coupled with densiometry, is currently being used in the Micromechanics of Solids Laboratory to assess the depth of corrosion pits in stainless steel coupons that are immersed in a solution whose active ingredient is thiosulphate for various lengths of time [49]. While radiography is sensitive enough for the measurement of pit depths, it has become abundantly clear during the course of this investigation that it is totally unsuited for crack initiation studies.

3.6.2 Transmission X-Ray Lang Camera Techniques. This procedure, whose application to the fatigue of single crystals is reported in [50], has a remote possibility of detecting crack initiation. The principle of this technique is that a collimated X-ray beam is directed towards the specimen and is diffracted by its crystallographic planes onto a slotted metal shield. The diffracted beam passing through this slot falls on a photographic plate while the combination of specimen and photographic plate are moved as a unit so that the entire stressed sample is scanned. Figure 9 shows the dislocation structure of a stressed single crystal of Silicone. While the procedure currently applies to single crystals, it has a small possibility of being applicable to coarse-grained polycrystals and hence to the study of crack initiation.

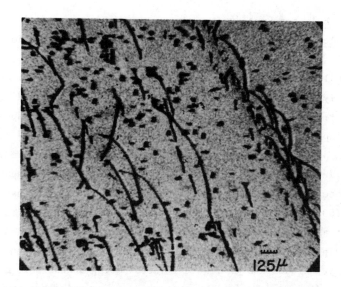

*Figure 9 - Dislocation Pattern of a Single Silicone Crystal under an
     External Stress of 2 Mpa*

3.6.3 <u>Ultrasonics</u>. In industry, the ultrasonic procedures, as
reviewed in [26] and [51], are currently considered to be the most reliable for in-the-field macrocrack detection. Their success if due to their
safety, versatility, and high sensitivity to sub-surface defects in metallic
and nonmetallic materials. All such techniques be they body or surface
wave reflected or attenuated, involve the beaming of pulsed elastic waves
into the specimen from an ultrasonic probe attached to the surface of
the specimen or component. The pulses scatter off a defect or crack and
onto a suitable positioned receiving probe. The position, size and shape
of the defect or crack can then be inferred from this received signal [52].
Since it has been stated in [53] that the presence of the plastic zone in
front of the crack reduces the accuracy of the method to the order of 1 mm,
it is unlikely that any currently available industrial equipment is sensitive enough to detect the micromechanics definition of initiation as given
above. The technique is, however, receiving a lot of development attention
at present and may, in the future, be able to detect initiation in laboratory as well as industrial situations.

3.7 Energy Emission Procedures

3.7.1 <u>Acoustic Emission (AE)</u>. This form of energy emission is in the
form of elastic waves, some of which possess ultrasonic frequencies, that
are internally generated when a specimen or component is subjected to a
sufficiently high stress state [26,51,54]. They are detected by means of
sensitive transducers that are located on the surface and respond to the
surface displacements that are produced by the acoustic waves. There are
two main types of transducers: resonant, which consist of piezo-electric
elements; and non-resonant that are air-gap capacitance transducers. The
latter detect a broader range of frequencies (0 to 45 MHz) than the piezo-
electric (50 Hz to 1.5 Mhz) but have much less sensitivity. The major
difficulty associated with AE measurements is the suppression of parasitic
noise originating from the testing machine or its surroundings. Electrical
filtering techniques and their use in attenuating undesirable noise is
extensively discussed in [55]. Apart from noise, this procedure is further
complicated by the multiple reflections and interferences occurring during
the propagation of the wave front prior to its arrival at the transducer.
It has been successfully applied to the study of fatigue crack growth
in steels, [56], and of inclusion fracture, [57], and stress corrosion
cracking, [58], in 7075 aluminium alloys. It was not successful in a preliminary investigation of crack initiation [59] in which O.F.H.C. copper
was fatigue failed. As is illustrated in Figure 10 the acoustic emission
bursts occurred in an erratic fashion which contributed to the outcome of
the experiment shown in Figure 11. This clearly shows that the standard
AE experiment is not capable of detecting crack initiation. Steps similar
to those discussed in [60] are at present being investigated.

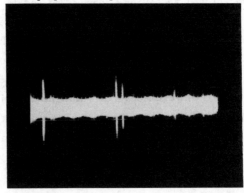

Figure 10 - An Illustration of AE Bursts During Fatigue

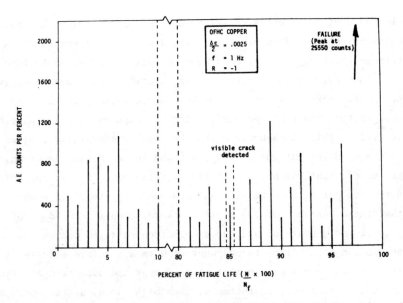

Figure 11 - *The Experimental Outcome Indicating the Nondetection of Initiation*

3.7.2 Thermal Energy Release. An estimation of both crack initiation and the cyclic endurance limit has recently been made by monitoring the surface temperature during fatigue loading of plain carbon steels [61,62]. Apart from the somewhat arbitrary temperature rise definitions of crack initiation, the method deserves consideration since it is envisaged that crack initiation, as defined here, will result in a change in the surface temperature after the fatigue crack has entered into its stage II phase. This method has the potential of not only being simple but also sensitive, responsive to local temperature changes and also adaptable to noncontacting temperature measuring techniques. As a further possibility, it may, in the future, be able to theoretically associate the formation of dislocation slip bands and PSBs to the measured temperature rise along with being able to associate the changes mentioned in Section 3.5 to these temperature changes.

3.7.3 Exoelectron Emission. Upon the plastic deformation of a ductile polycrystalline metal the brittle film of natural surface oxide that exists on most materials cracks open to reveal the fresh unoxidized surface of the underlying metal. Under ultraviolet illumination, photoelectrons are emitted more easily from the bare metal surface than from the surrounding oxide-coated material and, hence, the oxide layer rupture results in a

large increase of local emission referred to as photostimulated exoelectron emission (PSEE). In ambient air, the low energy (1 ev) electron is
almost immediately captured by an oxygen or nitrogen molecule to form a
molecular ion that may be attracted by an electric field to a suitably
designed probe where they become a direct manifestation, in the form of
picoamps, of the amount of new surface that is exposed during a particular
fatigue cycle. This atmospheric method has been successfully applied in
[63] to the study of fatigue crack growth and to the detection of which ho]
in a series of identical holes initiates a fatigue crack. In the spirit of
the initiation definition utilized in this paper, it is of interest to
note that the appearance of the first crack relieved the build up of
stress in its neighbourhood and the exoelectron emission decreased. Since
this method can be utilized in conjunction with a standard material testing
facility, it has a very good potential as far as the detection of crack
initiation is concerned. Furthermore, in a high vacuum system, such as
that described in [64] and [65], such electron emissions produce stable
images in a photoelectron microscope providing unique and direct observations of the rupture of the surface oxide layer. Such layers have been
shown to be in the order of 4 nm on steel and aluminium specimens whose
microcrack nucleation mechanism has been attributed to slip system controlled extrusions and intrusions. This method had already demonstrated
its ability to detect fatigue crack initiation and has, as a result, a
tremendous potential for successfully delineating fatigue crack initiation
in laboratory samples being tested in vacuum.

4.   CONCLUDING REMARKS
     This paper has highlighted both what is known concerning the micro-
mechanisms of fatigue crack initiation and the techniques that have a
possibility of nonarbitrarily assessing the number of cycles involved in
this very important phase of fatigue failure. It is conjectured that theor
alone will not be in a position to predict crack initiation without the
use of laboratory determined material crack initiation parameters which
will vary from material to material and hence a knowledge of the fundamenta
microprocesses taking place coupled with one or more precise experimental
procedures will in the future lead to a solution of this problem. When
this is accomplished to the satisfaction of the research and engineering

community then both reliability and damage tolerant design procedures will be radically improved. The motivation for solving this problem is certainly strong enough.

ACKNOWLEDGEMENTS

    The references cited in this paper are but a few of those which are available on the various subjects covered in this paper. Any omission is totally unintentional and should not be considered as a slight to the very excellent research currently taking place or undertaken in the past. This research was supported by the National Sciences and Engineering Research Counc of Canada under Grant No. A7525, whose financial assistance is gratefully acknowledged.

REFERENCES

1. PROVAN, J.W., "A Fatigue Reliability Distribution Based on Probabilistic Micromechanics", *Defects and Fracture*, edited by G.C. Sih and H. Zorski, Martinus Nijhoff Publishers, The Hague, 1982, pp. 63-69.
2. PROVAN, J.W. and THERIAULT, Y., "An Experimental Investigation of Fatigue Reliability Laws", *Defects, Fracture and Fatigue*, edited by G.C. Sih and J.W. Provan, Martinus Nijhoff Publishers, The Hague, 1983, pp. 423-432.
3. THERIAULT, Y., "An Experimental Investigation of Fatigue Reliability Laws", *Master's Thesis*, McGill University, 1983.
4. *1980 Annual Book of ASTM Standards*, 10, American National Standard, ANSI/ASTM E466-76, 1980, pp. 614-619.
5. PROVAN, J.W., "The Micromechanics Approach to the Fatigue Failure of Polycrystalline Metals", *Cavities and Cracks in Creep and Fatigue*, Chapter 6, edited by J. Gittus, Applied Science Publishers, London, 1981, pp. 197-242.
6. MBANUGO, C.C.I., "Stochastic Fatigue Crack Growth - An Experimental Study", *Ph. D. Thesis*, McGill University, 1979.
7. LAIRD, C., "Mechanisms and Theories of Fatigue", *Fatigue and Microstructure*, Material Science Seminar, St. Louis, 1978, American Society for Metals, 1979, pp. 149-203.
8. MITCHELL, M.R., "A Unified Predictive Technique ...", *SAE/SP-79/448*, Paper No. 790890, 1979.
9. MUGHRABI, H.,; ACKERMANN, F. and HERZ, K., "Persistent Slipbands in Fatigued Face-Centered and Body-Centered Cubic Metals", *Fatigue Mechanisms*, edited by J.T. Fong, ASTM STP 675, 1979, pp. 69-105.
10. KUHLMANN-WILSDORF, D. and LAIRD, C., "Dislocation Behaviour in Fatigue", *Mat. Sci. Eng.*, Vol. 27, 1977, pp. 137-156.
11. MUGHRABI, H. and WANG, R., "Cyclic Strain Localization ...", *Defects and Fracture*, edited by G.C. Sih and H. Zorski, Martinus Nijhoff Publishers, The Hague, 1982, pp. 15-28.
12. FINNEY, J.M. and LAIRD, C., "Strain Localization in Cyclic Deformation of Copper Single Crystals", *Phil. Mag.*, Vol. 31, 8th Series, 1975, pp. 339-366.

13. MUGHRABI, H., "A Model of High-Cycle Fatigue-Crack Initiation at Grain Boundaries by Persistent Slip Bands", *Defects, Fracture and Fatigue*, edited by G.C. Sih and J.W. Provan, Martinus Nighoff Publishers, The Hague, 1983, pp. 139-146.

14. POHL, K., MAYR, P. and MACHERAUCH, E., "Shape and Structure of Persistent Slip Bands in Iron Carbon Alloys", *Defects, Fracture and Fatigue*, edited by G.C. Sih and J.W. Provan, Martinus Nijhoff Publishers, The Hague, 1983, pp. 147-159.

15. POHL, K., MAYR, P. and MACHERAUCH, E., "Persistent Slip Bands in the Interior of a Fatigued Low Carbon Steel", *Scripta Met.*, Vol. 14, 1980, pp. 1167-1169.

16. LAIRD, C. and DUQUETTE, D.J., "Mechanisms of Fatigue Crack Nucleation", *Corrosion Fatigue: Chemistry, Mechanics and Microstructure*, edited by A.J. McEvily and R.W. Staehle, NACE, Houston, 1972, pp. 88.

17. COOPER, C.V. and FINE, M.E., "Fatigue Crack Initiation in Iron", *Defects, Fracture and Fatigue*, edited by G.C. Sih and J.W. Provan, Martinus Nijhoff Publishers, The Hague, 1983, pp. 183-194.

18. ESSMANN, U., GOESELE, U. and MUGHRABI, H., "A Model of Extrusions and Intrusions in Fatigued Metals, Part I", *Phil. Mag. A.*, Vol. 44, 1981, pp. 405.

19. KIM, W.H. and LAIRD, C., "Crack Nucleation and Stage I Propagation in High Strain Fatigue, I and II", *Acta. Met.*, Vol. 26, 1978, pp. 777-787 and pp. 789-799.

20. FORSYTH, P.J.E., *The Physical Basis of Metal Fatigue*, Blackie and Sons Ltd., London, 1969.

21. BUCK, O., MORRIS, W.L. and JAMES, M.R., "Remaining Life Prediction in the Microcrack Initiation Regime", *Fracture and Failure: Analyses, Mechanisms and Applications*, edited by P.P. Tung et al, American Society for Metals, 1981, pp. 55-64.

22. FINE, M.E. and RITCHIE, R.O., "Fatigue-Crack Initiation and Near-Threshold Crack Growth", *Fatigue and Microstructure*, Material Science Seminar, St. Louis, 1978, American Society for Metals, 1979, pp. 245-278.

23. BROWN, R. and SMITH G.C., "Mechanisms of Fatigue Crack Initiation in a Ti-6Al-4V Alloy", *Materials, Experimentation and Design in Fatigue*, edited by F. Sherratt and J.B. Sturgeon, IPC Business Press, Guildford, England, 1981, pp. 22-32.

24. FASH, J.W., SOCIE, D.F. and RUSSELL, E.S., "Fatigue Crack Initiation and Growth in Gray Cast Iron", *Materials, Experimentation and Design in Fatigue*, edited by F. Sherratt and J.B. Sturgeon, IPC Business Press, Guildford, England, 1981, pp. 40-51.

25. BUCK, O. and ALERS, G.A., "New Techniques for Detection and Monitoring of Fatigue Damage", *Fatigue and Microstructure*, Material Science Seminar, St. Louis, 1978, American Society for Metals, 1979, pp. 101-147.

26. GREEN, R.E., Jr., "Non-Destructive Methods for the Early Detection of Fatigue Damage in Aircraft Components", *AGARD Lecture Series*, No. 103, NATO, 1979, pp. 6-1 - 6-31.

27. HOEPPNER, D.W., "Model for Prediction of Fatigue Lives Based upon a Pitting Corrosion Fatigue Process", *Fatigue Mechanisms*, edited by J.T. Fong, ASTM STP 675, 1979, pp. 841-870.

28. JAVORNICZY, J., *Photoplasticity*, Elsevier Scientific Publications, 1974.

29. DUDDEDAR, T.D., "Application of Holography to Fracture Mechanics", *Experimental Mechanics*, 1969, pp. 281-285.

30. FRIESEM, A.A. and VEST, C.M., "Detection of Micro Fractures by Holographic Interferometry", *Appl. Opt.*, Vol. 8, 1979, pp. 1253-1254.

31. VEST, C.M., "Crack Detection", *Holographic Non-Destructive Testing*, edited by R.K. Erf, Academic Press, 1974.

32. ACHARD, L.-M., "On The Applicability of Holographic Interferometry to Assess J-Integral and C.O.D. Fracture Criteria", *Master's Thesis*, McGill University, 1982.

33. RICHARDS, C.E. and DEANS, W.F., "The Measurement of Crack Length and Load Using Strain Gauges", *The Measurement of Crack Length and Shape During Fracture and Fatigue*, edited by C.J. Beevers, EMAS, Chameleon Press, London, 980, pp. 28-68.

34. KAMATH, M.S. and NEAVES, H.J., "A Strain Gage Method for Detecting Ductile Crack Initiation", *Int. J. Fracture*, Vol. 14, 1978, pp. R199-R204.

35. WEI, R.P. and BRAZILL, R.L., "An AC Potential System for Crack Length Measurement", *The Measurement of Crack Length and Shape During Fracture and Fatigue*, edited by C.J. Beevers, EMAS, Chameleon Press, London, 1980, pp. 190-201.

36. HALLIDAY, M.D. and BEEVERS, C.J., "The DC Electric Potential Method for Crack Length Measurement", *The Measurement of Crack Length and Shape During Fracture and Fatigue*, edited by C.J. Beevers, EMAS, Chameleon Press, London, 1980, pp. 85-112.

37. PARIS, P.C. and HAYDEN, B.R., "A New System for Fatigue Crack Growth Measurement and Control", presented at the ASTM Symposium on Fatigue Crack Growth, Pittsburgh, October, 1979.

38. KIKUKAWA, M., JONO, M. and ADACHI, M., "Direct Observation and Mechanism of Fatigue Crack Propagation", *Fatigue Mechanisms*, edited by J.T. Fong, ASTM STP 675, 1979, pp. 234-253.

39. DAVIDSON, D.L. and LANKFORD, J., "Dynamic, Real-Time Fatigue Crack Propagation ...", *Fatigue Mechanisms*, edited by J.T. Fong, ASTM STP 675, 1979, pp. 277-284.

40. DAVIDSON, D.L., "The Study of Fatigue Mechanisms with Electron Channeling", *Fatigue Mechanisms*, edited by J.T. Fong, ASTM STP 675, 1979, pp. 254-275.

41. BEACHEM, C.D., "An Electron Fractographic Study ...", *Trans. American Society for Metals*, Vol. 56, 1963, pp. 318-326.

42. BEACHEM, C.D. and PELLOUX, R.M.N., "Electron Fractography - A Tool for the Study of Micromechanisms of Fracture Processes", *Fracture Toughness Testing and Its Applications*, ASTM STP 381, 1965, pp. 21-45.

43. BEACHEM, C.D. and MAYN, D.A., "Fracture by Microscopic Plastic Deformation Processes", *Electron Fractography*, ASTM STP 436, 1968, pp. 58-88.

44. WIEBE, W. and DAINTY, R.V., "Fractographic Determination of Fatigue Crack Growth Rates in Aircraft Components", *Canadian Aeronautics and Space Journal*, Vol. 27, 1981, pp. 107-117.

45. IMURA, T. and YAMAMOTO, A., "The Behaviour of Dislocations and the Formation of Wall Structures Observed by In Situ High Voltage Electron Microscopy", *Defects, Fracture and Fatigue*, edited by G.C. Sih and J.W. Provan, Martinus Nijhoff Publishers, The Hague, 1983, pp. 17-21.

46. IMURA, T., "Possibilities for In-Situ Observation of Basic Deformation Processes and of Other Dynamic Processes", *Kristall und Technik*, Vol. 14, 1979, pp. 1197-1208.

47. OHR, S.M., HORTON, J.A., CHANG, S.-J., "Direct Observations of Crack Tip Dislocation Behaviour During Tensile and Cyclic Deformation", *Defects, Fracture and Fatigue*, edited by G.C. Sih and G.W. Provan, Martinus Nijhoff Publishers, 1983, pp. 3-15.

48. RIE, K.-T. and KOHLER, W., "Effect of High Pressure Hydrogen on Low Cycle Fatigue", *Proc. of Int. Sym. on Low Cycle Fatigue Strength*, edited by K.-T. Rie and E. Haibach, DVM, Berlin, 1979, pp. 117-128.

49. RODRIGUEZ III, E.S., "A New Pitting Corrosion Failure Prediction Model", *Ph. D. Thesis*, McGill University, Montreal, 1984.

50. PROVAN, J.W. and GHONEM, H., "Microstress Distributions in Single Crystals of Silicon", *Canadian Metallurgical Quarterly*, Vol. 15, 1976, pp. 319-324.

51. GREEN, R.E., Jr. and DUKE, J.C., Jr., "Ultrasonic and Acoustic Emission Detection of Fatigue Damage", *Int. Advances in Nondestructive Testing*, Vol. 6, 1979, pp. 125-177.

52. KRAUTKRAMER, J. and KRAUTKRAMER, H., *Ultrasonic Testing of Materials*, Springer Verlag, Berlin, 1977.

53. CURRY, D.A. and MILNE, I., "The Detection and Measurement of Crack Growth during Ductile Fracture", *The Measurement of Crack Length and Shape During Fracture and Fatigue*, edited by C.J. Beevers, Chameleon Press, London, 1980, pp. 401-434.

54. HUTTON, P.H. and ORD, R.N., "Acoustic Emission", *Research Techniques and Nondestructive Testing*, edited by R.S. Sharpe, Vol. 1, 1970, pp. 1-30.

55. LINDLEY, T.C. and McINTYRE, P., "Application of Acoustic Emission to Crack Detection and Measurement", *The Measurement of Crack Length and Shape During Fracture and Fatigue*, edited by C.J. Beevers, Chameleon Press, London, 1980, pp. 285-344.

56. SINCLAIR, A.C.E., CONNORS, D.C. and FORMBY, C.L., "Acoustic Emission Analysis during Fatigue Crack Growth in Steel", *Materials Science and Engineering*, Vol. 28, 1977, pp. 263-273.

57. McBRIDE, S.L., MacLACHLAN, J.W. and PARADIS, B.P., "Acoustic Emission and Inclusion Fracture in 7075 Aluminum Alloys", *J. Nondestructive Evaluation*, Vol. 2, 1981, pp. 35-41.

58. DICKSON, J.I., MARTIN, P. and BAILON, J.-P., "The Study of Variations in Stress Corrosion Cracking Velocities in Aluminium Alloy 7075-T651 by Acoustic Emission", *Materials Science and Engineering*, Vol. 58, 1983, pp, L5-L8.

59. PROVAN, J.W., "Probabilistic Fatigue - Theory and Experimentation", *Int. J. Eng. Sci.*, submitted to.

60. GRAHAM, L.J. and ALERS, G.A., "Acoustic Emission in the Frequency Domain", *Monitoring Structural Integrity by Acoustic Emission*, ASTM STP 571, 1975, pp. 11-39.

61. DENGEL, D. and HARIG, H., "Estimation of the Fatigue Limit by Progressively Increasing Load Tests", *Fatigue in Eng. Matls. and Struct.*, Vol. 3, 1980, pp. 113-128.

62. HARIG, H. and WEBER, M., "Estimation of Crack Initiation in Plain Carbon Steels by Thermometric Methods", *Defects, Fracture and Fatigue*, edited by G.C. Sih and J.W. Provan, Martinus Nijhoff Publishers, The Hague, 1983, pp. 161-170.

63. HOENIG, S.A. et al, "Applications of Exoelectron Emission ...", *Testing for Prediction of Material Performance in Structures and Components*, ASTM STP 515, 1972, pp. 107-125.

64. BAXTER, W.T. and ROUZE, S.R., "The Effect of Oxide Thickness on Photostimulated Exoelectron Emission from Aluminum", *J. Appl. Phys.*, Vol. 49, 1978, pp. 4233-4237.

65. BAXTER, W.T., "The Role of Exoelectrons and Oxide Films in Fatigue Detection", *American Physical Society Topical Conference on Physics in the Automotive Industry*, AIP Conf. Proc. No. 66, 1980, pp. 107-119

MODELLING PROBLEMS IN CRACK TIP MECHANICS   *Waterloo, Ontario, Canada*
CFC10, University of Waterloo   *August 24-26, 1983*

MECHANICS OF NONLINEAR CRACK GROWTH: EFFECTS OF SPECIMEN SIZE
AND LOADING STEP

G.C. SIH and D.Y. TZOU
Institute of Fracture and Solid Mechanics
Lehigh University
Bethlehem, Pennsylvania, U.S.A.

1. INTRODUCTION

Nonlinear global load and displacement behaviour of metals is not
exclusively associated with yielding of material. It can arise in the
case of material damage by nonselfsimilar crack growth whereby the influence of the plastic deformation is negligible. The process of yielding
and crack growth are inherently interwoven such that their individual
effects cannot be readily separated either experimentally and/or analytically. This has led to frequent misinterpretation of the observed data
on material behaviour. Unless the combined influence of loading rate,
specimen size and material properties are understood, little progress can
be made to translate small specimen data to the design of larger size
structural components.

Since all materials possess nonhomogeneous microscopic structure,
their response in terms of yielding over a small region can be very different from those over a larger region. Plastic properties of metals collected
from a uniaxial test relies on the assumption of homogeneity such that
loading and unloading effects can be described on the uniaxial stress and
strain curve. Such an interpretation, however, loses its physical meaning
when applied to stress states in two- or three-dimensions, an obvious short-
coming of the current theory of plasticity. A more consistent modelling
of material damage at the microscopic level has been advanced in [1] for
metal with strain hardening and [2] for concrete with bilinear softening.
A pseudo-linear analysis was adopted where unloading is assumed to occur
elastically but permanent damage of the material is accounted for each loading
step by nonuniform alteration of the local moduli. Predictions were made
for the complete range of failure modes from brittle fracture to plastic
collapse.

It is generally recognized that metal fails by a process of excessive
deformation and fracture. Cracks are initiated and then spread slowly
prior to final termination [3]. This physical process can be conveniently
and consistently modelled by application of the strain energy density
criterion [4,5].

Assumed is that failure mode can be uniquely associated with the rate
at which energy is dissipated through a unit volume of material. This
quantity shall be referred to as the strain energy density function dW/dV
whose critical value $(dW/dV)_c$ or $(dW/dV)_c^*$ correspond to certain threshold
levels of material damage. The amount of energy released through a small
distance r can be measured by the strain energy density factor defined
as $S = r(dW/dV)$. Such a relation remains valid for all materials as S can
in general depend on all the space variables. When the material is damaged
by crack growth and yielding, parallel lines are obtained in the S versus
a plot when specimen sizes are varied with a being the full or half crack
length. The variations in loading step or rate and fracture toughness
can be represented by rotation of the S versus a lines. This approach pro
vides a means for predicting the load carrying capacity of a system and an
indication how specimen size trades off with loading rate for a given
material. Results are presented for a central crack panel that fails in
tension by a combination of yielding and fracture. The material undergoes
strain hardening.

2.   SUBCRITICAL AND CRITICAL MATERIAL DAMAGE

Material damage in metals occurs gradually by a process of yielding
and crack growth. A unique description of this process will be made by
focusing attention on an element of material ahead of the crack as shown
in Figure 1. The resolution of the continuum mechanics analysis is
denoted by the size of the core region r_0 which is small in comparison
with r, i.e., $r_0 \ll r$. The energy stored in the volume element can be
obtained from

$$\frac{dW}{dV} = \int_0^{\varepsilon_{ij}} \sigma_{ij} d\varepsilon_{ij} \ , \tag{1}$$

and decays with the distance r measured from the crack tip, Figure 1.
Equation (1) is valid for all materials undergo isothermal deformation
with σ_{ij} and ε_{ij} being the stress and strain components. Without loss in

generality, a strain energy density factor S can be defined as

$$\frac{dW}{dV} = \frac{S}{r} \cdot \tag{2}$$

For a linear elastic material, S can be computed from the theory of elasticity and the stress intensity factors k_j (j = 1,2,3) if attention is focused only on the onset of rapid crack extension. When energy is dissipated prior to crack growth, the calculation of S must account for it accordingly.

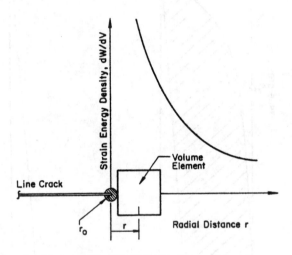

Figure 1 - Volume Element Next to Core Region with $r_0 \ll r$

2.1 Available Release Energy

Consider a uniaxial tension test where a segment of material within the gage length is assumed to be in a homogeneous state of stress. The various states of energy can then be obtained from a plot of the uniaxial stress and strain curve, a schematic representation of which is given in Figure 2. Suppose that unloading beyond the yield point e follows the line pp_1 parallel to oe, then the area $oepp_1$ represents the energy $(dW/dV)_p$ dissipated by yielding or plastic deformation. Fracture is assumed to occur when p reaches f. The energy available for release at that instant is

$$\left(\frac{dW}{dV}\right)_c^* = \left(\frac{dW}{dV}\right)_c - \left(\frac{dW}{dV}\right)_p \, , \tag{3}$$

which corresponds to the area f_1ff_2 while $(dW/dV)_c$ is the total area under the curve from o to f. In the absence of plasticity, all the energy is released elastically and $(dW/dV)_c^* = (dW/dV)_c$. *Subcritical crack growth is thus assumed to occur when the strain energy density dW/dV reaches the value* $(dW/dV)_c^*$.

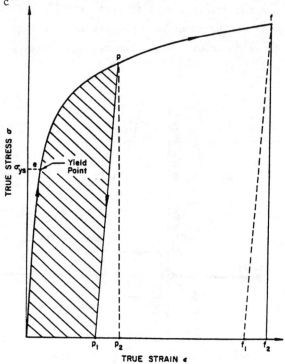

Figure 2 - Schematic Representation of an Elastic-Plastic Material

2.2 Sub-Critical Crack Growth

Once the crack starts to spread, its growth rate is assumed to follow the relation

$$\left(\frac{dW}{dV}\right)_c^* = \frac{S_1}{r_1} = \frac{S_2}{r_2} = \cdots = \frac{S_j}{r_j} = \cdots = \frac{S_c}{r_c} = \text{const.} \qquad (4)$$

Under the conditions that

$$S_1 < S_2 < \cdots < S_j < \cdots < S_c \; , \qquad (5)$$

$$r_1 < r_2 < \cdots < r_j < \cdots < r_c \; ,$$

then the crack growth rate is monotonically increasing and becomes unstable
when S and r attain their critical values S_c and r_c.

The direction of crack growth coincides with the relative minimum of
dW/dV or $(dW/dV)_{min}$ regardless of whether the material ahead of the crack
is yielded or not. This was verified in [6] by a plasticity analysis
showing that dW/dV attains a relative minimum in the region of gross yielding where the crack passes through.

The $(dW/dV)_{min}$ criterion on crack path prediction applies to all
materials and is independent of the constitutive relation.

2.3 Onset of Global Instability

Under increasing load, damage accumulates until a critical condition
is reached at which point the specimen separates into two. This event
can occur catastrophically or less dramatic depending on the rate of energy
released during the breakage of the last ligament of material. If the
release involves predominantly elastic energy, then the relation
$(dW/dV)_c = S_c/r_c$ applies such that S_c is directly associated with the
more common fracture toughness parameter K_{1c}:

$$S_c = \frac{(1+\nu)(1-2\nu)}{2\pi E} K_{1c}^2 , \tag{6}$$

where ν is the Poisson's ratio and E the Young's modulus.

Plastic deformation can still prevail off to the sides of the macrocrack even when final fracture is brittle. Because of the lack of precise
quantitative assessment of failure modes, it is not possible at this time
to clearly identify the various degrees of global instability by brittle
and/or ductile fracture. For instance, when plastic deformation dominates,
equation (4) should be modified as $(dW/dV)_c^* = S_c^*/r_c^*$ with $S_c^* < S_c$ and
$r_c^* < r_c$ remembering that failure modes can be altered by changing loading
rates and specimen sizes without changing the material. It would therefore
be less confusing to consider S_c or K_{1c} and S_c^* as material *behavioural*
parameters rather than material constants. There will of course be an
intermediate region as stated by equation (4), i.e., $(dW/dV)_c^* = S_c/r_c$.
This corresponds to global brittle fracture behaviour even though portion
of the energy has been dissipated by plastic deformations, equation (3).

3. NONLINEAR ELASTIC-PLASTIC STRESS ANALYSIS

The incremental theory of plasticity and the von Mises yield criterion are applied to solve the problem of a through crack of length 2a in a panel 2b x 2c x h as shown in Figure 3. Loading is applied beyond crack initiation and increased by steps of $\Delta\sigma$ as the crack spreads **slowly** in accordance with equations (4) and (5).

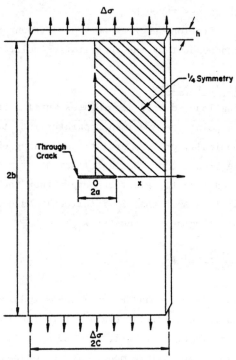

Figure 3 - Centre Cracked Panel under Uniform Tension

3.1 Material Types and Specimen Sizes

Referring to Figure 4, three different types of material will be considered. The values of yield strength σ_{ys}, fracture strength σ_{fs}, critical strain energy density function $(dW/dV)_c$, and available strain energy density function $(dW/dV)_c^*$ are given in Table 1. The curves in Figure 4 are modelled by the Ramberg-Osgood relation

$$\varepsilon = \begin{cases} \dfrac{\sigma}{E} ; & \sigma \leq \sigma_{ys} \\[2mm] \dfrac{1}{E}\left\{\sigma + \alpha\left[\left(\dfrac{\sigma}{\sigma_{ys}}\right)^\beta - 1\right]\sigma_{ys}\right\} \end{cases} , \qquad (7)$$

which is decomposed into 30 multilinear segments so as to achieve rapid convergence in the numerical analysis, i.e.,

$$\sigma_j = \sigma_{ys} + \left(\frac{\sigma_{fs} - \sigma_{ys}}{30}\right)(j-1) ,$$

$$\varepsilon_j = \frac{1}{E}\left\{\sigma_j + \alpha\left[\left(\frac{\sigma_j}{\sigma_{ys}}\right)^\beta - 1\right]\sigma_{ys}\right\} ,$$

(8)

with j = 1 to 31. In equations (7) and (8), $\alpha = .02$ and $\beta = 5.0$ are the strain hardening coefficients.

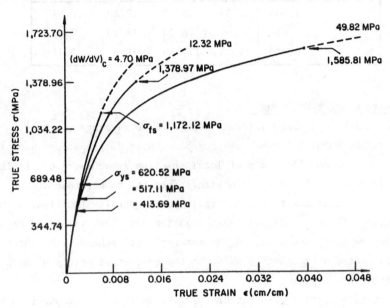

Figure 4 - True Stress and Strain Curve for Three Nonlinear Materials

Table 1 - Three Different Materials

Material Type	σ_{ys} (MPa)	σ_{fs} (MPa)	$(dW/dV)_c$ (MPa)	$(dW/dV)_c^*$ (MPa)
1	517.11	1,378.97	12.32	4.60
2	413.69	1,585.81	49.82	6.08
3	620.53	1,172.12	4.70	3.32

Specimen sizes are also varied in accordance with the dimensions in Table 2. Note that geometric similarities are preserved. Each size will be identified by the volume to surface ratio V/A.

Table 2 - Specimen Sizes

Specimen Type	V/A (cm)	Dimensions (cm)			
		a	b	c	h
1	1.104	2.54	25.40	17.20	2.54
2	2.931	7.62	76.20	38.10	7.62
3	6.840	17.78	177.80	88.90	17.78

3.2 Finite Element Grid Pattern

The twelve-node quadrilateral isoparametric elements are used in accordance with the PAPST [7] computer program. Cubic displacement shape functions are employed. Accuracy of local solution is achieved by placing four side nodes at 1/9 and 4/9 distance from the corner node at the crack tip. The 1/r singularity of the strain energy density function is thus preserved at these boundaries. Although the numerical values of dW/dV at the nodes may vary depending on the interpolation scheme, this uncertainty is circumvented by evaluating dW/dV at the quadrature points of each element with contour plots.

The finite element grid pattern for one quarter of the panel in Figure is displayed in Figure 5. All specimens are first loaded up to a stress level[1] of 413.69 MPa and then subjected to incremental loading. Three different cases with constant load steps are considered and they are identified with $\Delta\sigma$ = 155.13, 189.61 and 241.32 MPa.

---

[1] This is not the load at crack initiation which depends on specimen size, a topic that will be discussed in future communication.

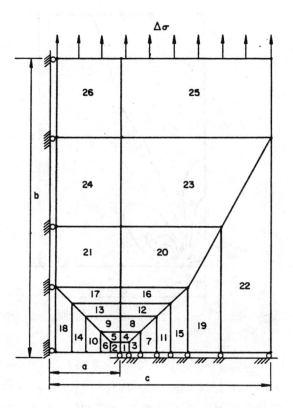

*Figure 5 - Finite Element Grid Pattern for Elastic-Plastic Stress
Analysis with One-Quarter Symmetry*

4. DISCUSSION OF RESULTS

One of the main objectives of this work is to illustrate how specimen
size, loading rate and material type effects can be assessed quantitatively
while the crack grows slowly through the yielded material. A typical display of the effective stress contours labelled from 1 to 8, with increasing
yield intensity, is exhibited in Figure 6. Material elements enclosed by
contours, say 6, 7 and 8 are yielded prior to crack growth. The energy
released by a segment of macrocrack growth therefore is only $(dW/dV)^*_c$
since $(dW/dV)_p$ in equation (3) has already been dissipated by plastic
deformation. The corresponding strain energy density factor S is thus
evaluated as a function of crack length.

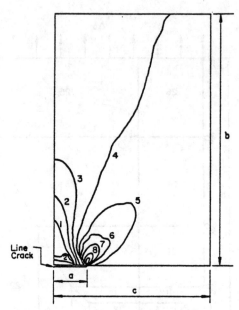

Figure 6 - Effective Stress Contours Exhibiting Yielding Pattern
at Crack Front

4.1 Specimen Size Effect

When specimen size increases, the ratio of volume to surface ratio
increases accordingly, Table 2. For a constant loading step of $\Delta\sigma$ = 241.3
MPa and material with $(dW/dV)_c$ = 12.32 MPa, the values of S versus a for
the three different V/A ratios are shown in Table 3. These results when
plotted graphically appear as parallel lines in Figure 7. For a given
material or constant S_c value, the smaller specimen with the lower V/A
ratio can sustain more subcritical crack growth. This explains the tendency for thicker or larger specimens to behave in a brittle fashion.

4.2 Variation in Loading Steps

Suppose that the loading steps are varied while the specimen size
and material type are kept constant with V/A = 1.104 cm and $(dW/dV)_c$ =
12.32 MPa. It is seen that the S versus a lines rotate in a counterclockwise direction as the loading step is increased from $\Delta\sigma$ = 155.13 to 241.32
MPa, Figure 8. Small loading step is seen to enhance or prolong slow
crack growth while large loading step favours more brittle type of failure

Table 3 - Effect of Specimen Size on Crack Growth for $\Delta\sigma = 241.32$ MPa and $(dW/dV)_c = 12.32$ MPa

V/A (cm)	a (cm)	S (x 10^3N/m)
1.104	2.540	1.057
	2.563	2.480
	2.617	5.650
2.931	7.620	4.0450
	7.708	10.388
	7.934	26.112
6.840	17.780	7.952
	17.953	19.168
	18.37	45.964

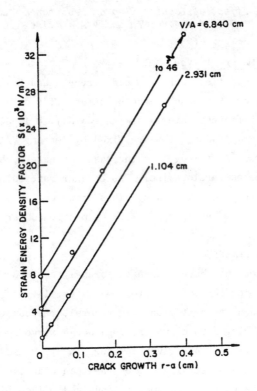

Figure 7 - Size Effect on Resistance Curve for $\Delta\sigma = 241.32$ MPa and $(dW/dV)_c^* = 4.60$ MPa

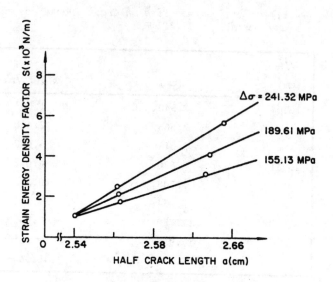

Figure 8 - Effect of Loading Step on Resistance Curve for
$(dW/dV)^*_c = 4.60$ *MPa and V/A = 1.104*

4.3 Change in Material Type

Material properties such as yield strength and fracture toughness can
also alter the crack growth characteristics. The data in Table 5 correspond to three different materials with $(dW/dV)_c$ = 12.32, 49.82 and 4.70 MPa
Specimen size and loading step are fixed by letting V/A = 1.104 cm and
$\Delta\sigma$ = 241.32 MPa. As $(dW/dV)^*_c$ is increased, Figure 9 shows that the S and
a lines rotate counterclockwise. The tougher material offers better resistance to failure by crack growth. This, however, is governed by a combination of loading and specimen size as mentioned earlier.

5. CONCLUDING REMARKS

A methodology has been presented to analyze the combined influence of
specimen size, loading rate or step and material type by application of the
strain energy density criterion. In general, the condition dS/da = const
is preserved during slow crack growth and can serve as a useful tool for
optimizing material with a given design. In a S versus a plot, all results
can be represented by straight lines than can translate and/or rotate
depending on the physical parameters involved. The wide range of failure
modes from brittle fracture to plastic collapse can now be assessed

quantitatively. This work is in contrast to that presented in [8] that considered crack growth as a series of crack initiation processes. The cracked specimen was unloaded globally for each increment of crack growth. The results in Figures 7 to 9 pertain to sustained global loading while unloading may occur only locally for elements near the crack. To be noted is that the condition dS/da = const remains unchanged in both of these cases.

Table 4 - Crack Growth Influenced by Change in Loading Step with V/A = 1.104 cm and $(dW/dV)_c$ = 12.32 MPa

$\Delta\sigma$ (MPa)	a (cm)	S (x 10^3N/m)
241.32	2.540	1.057
	2.563	2.480
	2.617	5.650
189.61	2.540	1.057
	2.563	1.425
	2.608	2.068
155.13	2.540	1.057
	2.563	1.963
	2.606	3.121

Table 5 - Different Material Properties for V/A = 1.104 cm and $\Delta\sigma$ = 241.32 MPa

$(dW/dV)_c^*$ (MPa)	a (cm)	S (x 10^3N/m)
4.60	2.540	1.057
	2.563	2.480
	2.617	5.650
6.08	2.540	1.277
	2.561	4.134
	2.608	10.312
3.32	2.540	0.963
	2.569	1.827
	2.624	3.611

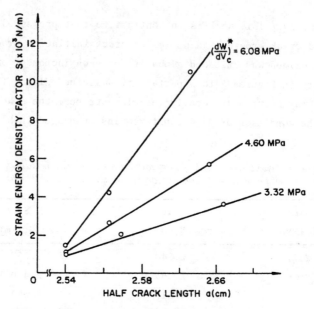

Figure 9 - Crack Growth Resistance Curves for Different Fracture Toughness with $\Delta\sigma = 241.32$ MPa and $V/A = 1.104$

Several shortcomings still prevail in the analysis of nonlinear crack growth. First, the theory of plasticity is not adequate as the von Mises yield criterion considers only the distortional component of the strain energy density function. The concept of loading and unloading cannot be effectively carried over from the uniaxial tensile test to the multi-axial stress field near the crack. The complete material damage process should be addressed from initiation to final termination. Specimen with different sizes will initiate crack growth at different load levels. These and other inadequacies in the present investigation will be overcome in the future.

REFERENCES

1. SIH, G.C. and MATIC, P., "A Pseudo-Linear Analysis of Yielding and Crack Growth: Strain Energy Density Criterion", *Proceedings on Defects, Fracture and Fatigue*, edited by G.C. Sih and J.W. Provan, Martinus Nijhoff Publishers, The Netherlands, 1983, pp. 223-232.
2. CARPINTERI, A. and SIH, G.C., "Damage Accumulation and Crack Growth in Bilinear Materials with Softening: Application of Strain Energy Density Theory", *J. of Theoretical and Appl. Fracture Mechanics*, Vol. 1, No. 2, 1984.
3. SIH, G.C., "Mechanics of Ductile Fracture", *Proc. Int. Conf. on Fracture Mechanics Technology*, edited by G.C. Sih and C.L. Chow, Sijthof Noordhoff Int. Publishers, The Netherlands, 1977, pp. 767-784.

4. SIH, G.C., "Prediction of Crack Growth Characteristics", *Proc. Sym. on Absorbed Specific Energy and/or Strain Energy Density Criterion,* edited by G.C. Sih, E. Czoboly and F. Gillemot, Akadémiai Kiadó, Budapest, 1980, pp. 3-16.

5. SIH, G.C., "A Special Theory of Crack Propagation", *Mechanics of Fracture, Vol. I: Methods of Analysis and Solutions of Crack Problems,* edited by G.C. Sih, Noordhoff International Publishing, The Netherlands, 1973, pp. 21-45.

6. SIH, G.C. and MADENCI, E., "Fracture Initiation under Gross Yielding: Strain Energy Density Criterion", *J. of Engineering Fracture Mechanics,* Vol. 78, No. 3, 1983, pp. 667-677.

7. GIFFORD, L.N. and HILTON, P.D., *Preliminary Documentation of PAPST Nonlinear Fracture and Stress Analysis by Finite Elements,* David W. Taylor Naval Ship Research and Development Center, Bethesda, Maryland, January, 1981.

8. SIH, G.C. and MADENCI, E., "Crack Growth Resistance Characterized by the Strain Energy Density Function", *J. of Engineering Fracture Mechanics,* Vol. 16, (in press).

Part 3
Session Papers

MODELLING PROBLEMS IN CRACK TIP MECHANICS   *Waterloo, Ontario, Canada*
CFC10, University of Waterloo   *August 24-26, 1983*

STRAIN RATE DEPENDENCE OF FRACTURE TOUGHNESS (J_{IC}) AND DUCTILITY

M.R. BAYOUMI and M.N. BASSIM
Department of Mechanical Engineering
University of Manitoba
Winnipeg, Manitoba, Canada

1. INTRODUCTION

In most cases in engineering practice, fracture initiation in metals
and alloys is experimentally determined under slow loading conditions.
However for materials with strong temperature and strain rate dependence,
the fracture toughness usually decreases with decreasing temperature and
increasing loading rates. It is thus important to obtain a quantitative
correlation between ductility and fracture toughness of these materials as
a function of temperature and through a wide range of loading rates. Experimental data in these ranges reveals a transition region from high to low
values of fracture toughness.

A study of the variation of fracture toughness J_{IC} and ductility, measured under both tensile loading and biaxial plane strain (bulge) loading
of AISI 1045 steel, in the transition temperature range was carried out [1].
The temperature range used showed changes in behaviour from primarily linear
elastic to elastic-plastic for this steel. A model relating fracture toughness, expressed as J_{IC} and bulge ductility for materials exhibiting linear
elastic behaviour at low temperature and elastic-plastic at higher temperatures was reported [2], while a model based on ductile fracture mechanisms
involving void nucleation followed by cavity growth and void coalescence
was developed to relate the fracture toughness parameter J_{IC} with temperature [3].

Determination of the fracture toughness of materials under high loading
rates, in the dynamic regime, has received much less attention than in the
case of quasi-static loading. Presently, there is an increased interest
in dynamic fracture studies motivated in part by problems associated with
transport and energy applications. The difficulties in studying dynamic
fracture behaviour of materials lies in both analytical and experimental
aspects.

In the analysis of dynamic fracture $\dot{K}_I$ defined as

$$\dot{K}_I = \frac{K_{IC}}{t_c} ,$$

(1)

where t_c is the time interval from the start of loading to the point where the crack starts to propagate, is frequently used to characterize how fast the crack tip region is loaded. The loading rate parameter $\dot{K}_I$ may be changed over several orders of magnitude, mainly due to changes in the loading time t_c.

The most common experimental technique to study dynamic fracture is the Instrumented Charpy test. Difficulties in understanding the inertia forces and the wave mechanics in the Charpy specimen limit the interpretation of the resulting data from this test. Another testing alternative for studying fracture at higher loading rates is based on the Split Hopkinson Pressure Bar system [4,5]. An experimental method was described for measuring the fracture properties of metals and alloys over a wide range of loading rates, which can cover over six orders of magnitude in $\dot{K}_I$ [5]

$$(1 \text{ MPa}\sqrt{m} \text{ s}^{-1} \le \dot{K}_I \le 10^6 \text{ MPa}\sqrt{m} \text{ s}^{-1}).$$

Studies in dynamic fracture have been limited to analysis based on linear elastic fracture mechanics concepts. In many engineering materials, such as low and medium strength steels, fracture is analyzed in terms of elastic-plastic behaviour using the J-integral. Recently wedge loaded compact tension specimens (WLCT) were adopted for quasi-static and dynamic determination of J_{IC} using a special arrangement of the Split Hopkinson Pressure Bar [6]. The analysis of the results was based on the J_{IC} expression as:

$$J_{IC} = \frac{kA}{Bb} ,$$

(2)

where k is a constant which was found to be about equal to 1.0, A is the energy under load-displacement curve up to the point of onset of crack propagation, b is the remaining ligament and B is the specimen thickness. The value of k equal to 1.0 was confirmed using stretch zone measurements [6]. The values of equivalent K_{IC} for both quasi-static and dynamic tests were estimated using the equation

$$K_{IC} = \left[\frac{E \, J_{IC}}{(1-\nu^2)} \right]^{1/2} ,$$

(3)

where E is Young's modulus and ν is Poisson's ratio. A summary of the results is shown in Table 1.

Table 1 - Results of WLCT Specimens under Quasi-Static and Dynamic Testing [6]

Type of Testing	$J_{IC} = A/Bb$ KJ/m^2	Equivalent K_{IC} MPa $\sqrt{m}$	Critical Time t_c µs	$\dot{K}_I = K_{IC}/t$ MPa $\sqrt{m}$ s^{-1}
Quasi-Static	65.7	119.4	3.12×10^8	0.38
	67.5	122.67	3×10^8	0.4
	62.5	112.9	2×10^8	0.6
Dynamic	28.1	46	30	1.5×10^6
	22.5	36.9	30	1.23×10^6
	30.1	49.1	27	1.8×10^6
	31.6	51.6	20	2.5×10^6
	33	54	25	2.16×10^6

In this present study, a model relating fracture toughness, expressed as J_{IC} and ductility with strain rate expressed as $\dot{\varepsilon}$ or its equivalent loading rate $\dot{K}_I$ is developed. Such a model is applicable to metals and alloys which show a strong variation in mechanical properties as well as fracture behaviour with strain rate (or loading rate).

2. THE MODEL

Because of the strong dependence of fracture toughness on ductility, particularly in the elastic-plastic regime, the approach for determination of the strain rate dependence of J_{IC} relies on examination of the micro-structural events which control ductile fracture. Previous investigations have shown that ductile fracture involves two successive damage processes, namely the nucleation of cavities at inclusions and grain boundaries followed by cavity growth and void coalescence.

Following the same approach as reported in [3] it is possible to develop a model for fracture toughness J_{IC} and ductility dependence with strain rate. Namely, it is assumed that ductile mechanisms such as nucleation of voids at inclusions or grain boundaries takes place at low strain rates when either the inclusion or the matrix inclusion interface is subjected to critical normal stress σ_{rr}, this stress can be written as:

$$\sigma_{rr}(\dot{\varepsilon}) = Y\,(\bar{\varepsilon}^p,\dot{\varepsilon}) + \sigma_m(\dot{\varepsilon}) \ , \tag{4}$$

where $Y\,(\bar{\varepsilon}^p,\dot{\varepsilon})$ is the flow stress in the region of the inclusion, $\sigma_m(\dot{\varepsilon})$ is the mean stress for void nucleation and $\bar{\varepsilon}^p$ is the mean plastic strain.

Because $Y(\bar{\varepsilon}^p,\dot{\varepsilon})$ has a strong strain rate dependence, equation (4) indicates that the critical mean stress required for void nucleation varies with strain rate as

$$\sigma_m(\dot{\varepsilon}) = \sigma_{rr}(\dot{\varepsilon}) - Y(\bar{\varepsilon}^p,\dot{\varepsilon}) \ , \tag{5}$$

this equation assumes that σ_{rr} varies moderately with increasing strain rate (like Young's modulus).

For a material obeying the equation

$$\bar{\sigma} = K\,\bar{\varepsilon}^n = 3\,\sigma_m\,f(\alpha,\beta) \ , \tag{6}$$

where $\bar{\sigma}$ is the effective stress, $\bar{\varepsilon}$ is the effective strain, n is the strain hardening exponent and $f(\alpha,\beta)$ is a function of the stress state, the strain for void nucleation $\bar{\varepsilon}_{i,\alpha,\beta}$ which is obtained by equating $\bar{\varepsilon} = \bar{\varepsilon}_{i,\alpha,\beta}$ in equation (6) becomes a function of σ_m only if the stress state (α,β) remains the same, where $\alpha = \sigma_2/\sigma_1$, $\beta = \sigma_3/\sigma_1$, and σ_1, σ_2, σ_3 are the principal stress components. Void coalescence strain $\bar{\varepsilon}_{c,\alpha,\beta}$ can be expressed as:

$$\bar{\varepsilon}_{c,\alpha,\beta} = \bar{\varepsilon}_{c,\alpha,\beta}(\nu_f,n) \ . \tag{7}$$

Equation (7) indicates that, for a given material and constant stress state, the strain required for void coalescence strain is independent of strain rate. The principal influence on $\bar{\varepsilon}_{c,\alpha,\beta}$ is the volume fraction of the voids ν_f.

The two main components of the fracture strain $\bar{\varepsilon}_{F,\alpha,\beta}$ are the nucleation strain $\bar{\varepsilon}_{i,\alpha,\beta}$ and the coalescence strain $\bar{\varepsilon}_{c,\alpha,\beta}$. These two strains are additive, thus

$$\bar{\varepsilon}_{F,\alpha,\beta} = \bar{\varepsilon}_{i,\alpha,\beta} + \bar{\varepsilon}_{c,\alpha,\beta} \ . \tag{8}$$

Using equations (5), (6) and (7), the strain rate dependence of the fracture strain $\bar{\varepsilon}_{F,\alpha,\beta}$ can be written as:

$$\bar{\varepsilon}_{F,\alpha,\beta}(\dot{\varepsilon}) = \bar{\varepsilon}_{i,\alpha,\beta}(\sigma_{rr}(\dot{\varepsilon}) - (\dot{\varepsilon},n)) + \bar{\varepsilon}_{c,\alpha,\beta}(\nu_f,n) \ , \tag{9}$$

where $\bar{\varepsilon}_{F,\alpha,\beta}(\dot{\varepsilon})$ is the fracture strain as a function of strain rate, $Y(\dot{\varepsilon},n)$ is the flow stress as a function of strain rate which can be expressed as:

$$Y(\dot{\varepsilon}) = Y_{static} \left(\frac{\dot{\varepsilon}}{\dot{\varepsilon}_{static}} \right)^m_{Temperature,\ structure} \qquad (10)$$

where m is the strain rate sensitivity, defined as:

$$m = \frac{\partial\ Log\ \sigma}{\partial\ Log\ \dot{\varepsilon}} \qquad (11)$$

and ranges from 0.01 - 0.02 for most steels and high strength aluminium alloys, and $\sigma_{rr}(\dot{\varepsilon})$ is the interface stress normal to the inclusion-matrix interface as a function of strain rate. It is assumed that σ_{rr} varies with strain rate as:

$$\sigma_{rr}(\dot{\varepsilon}) = (\sigma_{rr})_{static} + \alpha_0\dot{\varepsilon}\ , \qquad (12)$$

where α_0 is a constant much smaller than one.

Examination of equation (9) shows that the dependence of fracture strain on strain rate is similar to the strain for void nucleation since the void coalescence strain is only dependent on the volume fraction of voids as well as the stress state. Thus, a plot of $\bar{\varepsilon}_{F,\alpha,\beta}$ versus strain rate should be parallel to that of $\bar{\varepsilon}_{i,\alpha,\beta}$ versus $\dot{\varepsilon}$ displaced upwards by an amount corresponding to $\bar{\varepsilon}_{c,\alpha,\beta}$.

Following [2,3], the relationship between fracture toughness J_{IC} and ductility $\bar{\varepsilon}_{F,\alpha,\beta}$ is:

(a) for low strain rates (elastic-plastic behaviour)

$$J_{IC}(\dot{\varepsilon}) = K_0 \frac{m_0\sigma_0}{c} L_p^* \bar{\varepsilon}_{F,\alpha,\beta}(\dot{\varepsilon})\ , \qquad (13)$$

(b) for high strain rates (predominantly linear elastic behaviour)

$$J_{IC} = S\rho^* \bar{\varepsilon}_{F,\alpha,\beta}^2(\dot{\varepsilon}) \cdot f(E,K,n,\bar{\varepsilon}_y) + 2\sigma_f\pi L_e^*(1-\nu^2)/E\ , \qquad (14)$$

where S is the shape factor characterizing the geometry of the plastic zone and is taken as equal to 1.0, ρ^* is Neuber's microsupport effect constant $\simeq$ 0.025 mm (0.001), $\bar{\varepsilon}_y$ is the yield strain equal to σ_y/E, σ_f is the cleavage fracture stress, L_e^* is a characteristic distance which depends on the microstructure of the material, C, K_0 are constants, m_0 is a constraint factor which depends on the material and testing conditions, σ_0 is the flow stress, and L_p^* is a characteristic length corresponding to the distance from the crack tip to a point where the strain reaches a critical value at crack initiation.

Substituting equation (9) into equations (13) and (14) gives the strain rate dependence of J_{IC}, namely:

(a) for low strain rates

$$J_{IC}(\dot{\varepsilon}) = K_0 \frac{m_0 \sigma_0}{c} L_p^* [\bar{\varepsilon}_{i,\alpha,\beta} [\sigma_{rr}(\dot{\varepsilon}) - Y(\dot{\varepsilon},n)] + \bar{\varepsilon}_{c,\alpha,\beta}(\nu_f,n)] , \qquad (15)$$

(b) for high strain rates

$$J_{IC}(\dot{\varepsilon}) = S\rho^* [\bar{\varepsilon}_{i,\alpha,\beta} [\sigma_{rr}(\dot{\varepsilon}) - Y(\dot{\varepsilon},n)] + \bar{\varepsilon}_{c,\alpha,\beta}(\nu_f,n)]^2 \cdot$$

$$f(E,K,n,\bar{\varepsilon}_y) + 2\sigma_f \pi L_e^* (1-\nu^2)/E . \qquad (16)$$

In equations (15) and (16), it can be observed that the dependence of J_{IC} on strain rate follows that of the void nucleation step in the fracture process. Thus, it is necessary to determine the variation of the flow stress $Y(\dot{\varepsilon},n)$ and the interface stress $\sigma_{rr}(\dot{\varepsilon})$ with strain rate to obtain relationships of the variation of J_{IC} with strain rate in both linear elastic and elastic-plastic regimes for a given material and stress state.

3. APPLICATION OF THE MODEL

The data from [6] which are given in Table 1 are used in this section. In equation (10), Y_{static} is evaluated to be 450 MPa [1] at strain rate $\dot{\varepsilon}$ corresponding to 10^{-3} s^{-1} and m is taken to be 0.018, while in equation (12), $(\sigma_{rr})_{static}$ is estimated to be 866 MPa for Fe_3C partiales in 1045 steel at $\dot{\varepsilon} = 10^{-3}$ s^{-1}, also α_0 in equation (12) is assumed to be 0.002, thus equations (10) and (12) becomes:

$$Y(\dot{\varepsilon}) = 450 \left(\frac{\dot{\varepsilon}}{10^{-3}}\right)^{0.018} MPa , \qquad (17)$$

and

$$\sigma_{rr}(\dot{\varepsilon}) = 866 + 0.002 \dot{\varepsilon} \text{ MPa} . \qquad (18)$$

Using equations (17) and (18) it is possible to calculate the ratio of

$$\frac{\bar{\varepsilon}_{i,\alpha,\beta}(\dot{\varepsilon})}{\bar{\varepsilon}_{i,\alpha,\beta}(\dot{\varepsilon} = 10^{-3} s^{-1})}$$

as a function of strain rate, Figure 1 shows the variation of this ratio with Log $\dot{\epsilon}$. It is noticed that it consists of two straight lines which intersect at Log $\dot{\epsilon} \simeq 0.3$.

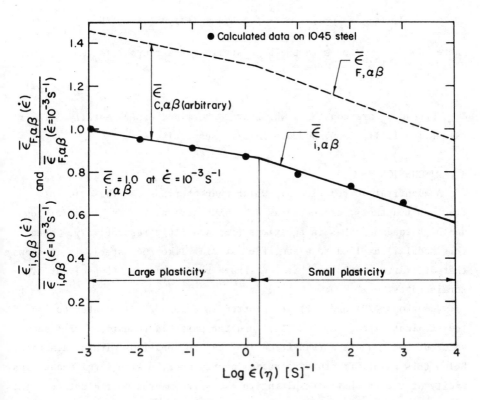

Figure 1 - *Strain Rate Dependence of $\bar{\epsilon}_{i,\alpha,\beta}$, expressed as $\bar{\epsilon}_{i,\alpha,\beta}(\dot{\epsilon})/\bar{\epsilon}_{i,\alpha,\beta}$ ($\dot{\epsilon} = 10^{-3}s^{-1}$), and $\bar{\epsilon}_{F,\alpha,\beta}$, expressed as $\bar{\epsilon}_{F,\alpha,\beta}(\dot{\epsilon})/\bar{\epsilon}_{F,\alpha,\beta}$ ($\dot{\epsilon} = 10^{-3}s^{-1}$.) for AISI 1045 Steel. $\bar{\epsilon}_{c,\alpha,\beta}$ denotes an arbitrary value for $\bar{\epsilon}_{F,\alpha,\beta}$.*

Next, it is now possible to determine the variation of J_{IC} with $\dot{\epsilon}$ in both linear elastic and elastic-plastic regimes for this material

    (a) for low strain rates

$$J_{IC}(\dot{\epsilon}) = K_0 \frac{m_0 \sigma_0}{c} L_p^* \zeta_1 \cdot \eta , \qquad (19)$$

where $\zeta_1 = \bar{\epsilon}_{F,\alpha,\beta}/\text{Log } \dot{\epsilon}$ in the low strain rates range, and $\eta = \text{Log } \dot{\epsilon}$. In general form equation (19) is:

$$J_{IC}(\dot{\epsilon}) = C_{ep} \cdot \eta , \qquad (20)$$

where C_{ep} is a constant which depends on the mechanical properties and microstructure of the steel and is determined experimentally.

(b) for high strain rates

$$J_{IC}(\dot{\varepsilon}) = S\rho^* \zeta_2^2 \cdot \eta^2 \cdot f(E,K,n,\bar{\varepsilon}_y) + 2\sigma_f \pi L_e^* (1-\nu^2)/E , \qquad (21)$$

where $\zeta_2 = \bar{\varepsilon}_{F,\alpha,\beta}/Log \, \dot{\varepsilon}$ in the high strain rates range. In general form

$$J_{IC}(\dot{\varepsilon}) = C_0 + C_e \cdot \eta^2 , \qquad (22)$$

C_{ep}, C_0 and C_e are constants which are determined experimentally. Their values are 1.111, 22.5 and 20.0 KJ·S/m^2 respectively.

4. DISCUSSION

A quantitative relationship which characterizes the variation of fracture toughness, expressed as J_{IC}, with strain rate is given for low and high loading rates in equations (15) and (16), respectively. When this model is applied to a specific material like 1045 steel, it is shown that J_{IC} varies with $Log \, \dot{\varepsilon}$ at low strain rates and with $(Log \, \dot{\varepsilon})^2$ at high strain rates.

Equations (20) and (22) are plotted in Figure 2. The intersection of the straight line (equation (20)) and the parabola (equation (22)) shows a value of $Log \, \dot{\varepsilon}$ where a transition in fracture behaviour with strain rate. Below this value, fracture is predominantly ductile with significant plasticity at the crack tip resulting in excessive macroscopic blunting. This is evident from observations of the fractured surface using scanning electron microscopy and measurement of the stretch zone width. Above this value of strain rate, the fracture behaviour is mostly elastic and the corresponding stretch zone ahead of the crack tip is very limited.

ACKNOWLEDGEMENT

The support of the Natural Sciences and Engineering Research Council of Canada is acknowledged.

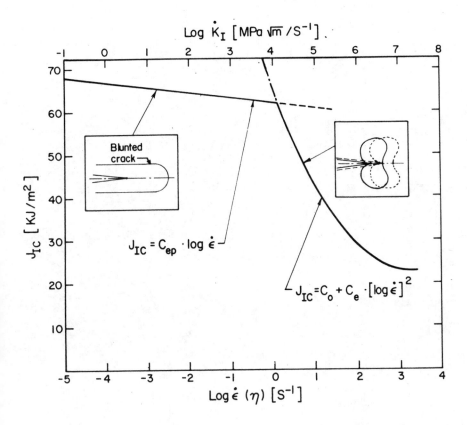

Figure 2 - Relationship between J_{IC} and Strain Rate

REFERENCES

1. BAYOUMI, M.R. and BASSIM, M.N., *Eng. Fracture Mechanics*, Vol. 18, No. 3, 1983, pp. 679-691.
2. BAYOUMI, M.R. and BASSIM, M.N., *Int. J. of Fracture*, Vol. 23, No. 1, 1983, pp. 71-79.
3. BAYOUMI, M.R. and BASSIM, M.N., *Int. J. of Fracture*, in press.
4. COSTIN, L.S. and DUFFY, J., *ASME, J. of Engineering Materials and Technology*, Vol. 101, 1979, pp. 258-264.
5. KLEPACZKO, J.R., *ASME, J. of Engineering Materials and Technology*, Vol. 104, 1982, pp. 29-35.
6. BAYOUMI, M.R., KLEPACZKO, J. and BASSIM, M.N., *J. of Testing and Evaluation*, in press.

A NUMERICAL STUDY OF CRACK TIP PARAMETERS IN CREEPING SOLIDS

S.B. BINER, D.S. WILKINSON   D. WATT
Institute for Materials Research   Department of Engineering Materials
McMaster University   University of Windsor
Hamilton, Ontario, Canada   Windsor, Ontario

1. INTRODUCTION

There is a clear need for reliable procedures capable of assessing the integrity of structural components which operate in the creep regime. However, due largely to a lack of understanding of the parameters which control crack growth during creep, present procedures often require that a component be retired from service while a significant portion of its life may still remain. The crack tip parameters commonly used when attempting to correlate crack growth rate are the stress intensity factor K [1], the net section stress [2], the crack tip opening displacement rate [3,4] and C* [5,6], the rate dependent J-integral. Experimental studies [1-7] have shown that while crack growth rates obtained from a single geometry may correlate well with some of these parameters, correlation for a range of geometries is generally poor (although C* does appear to show some promise in this regard).

The initial response of a cracked specimen, upon application of a load, and for some time thereafter is elastic. During this initial period, creep strains are negligible everywhere (i.e., less than the elastic strains), except in the immediate vicinity of the crack tip. This is of course analogous to the small-scale yielding situation at low temperatures, and K provides an adequate description of the crack tip stress and strain field.

The long-time response of a creeping structure is generally modelled using a power-law constitutive equation for the material, i.e.,

$$\dot{\varepsilon}_e = B\sigma_e^{\,n} ,\tag{1}$$

where $\dot{\varepsilon}_e$ and σ_e are the von Mises equivalent strain rate and stress respectively. B and n are material constants (B being strongly temperature dependent). The analytical studies of Rice and Riedel [7] (supported by numerical analyses [8,9]) have shown that once a steady-state has been achieved, the stress and strain fields near the crack tip have an HRR singularity with the path independent integral C* as the amplitude, i.e.,

$$\sigma_{ij} = \left(\frac{C^*}{B\, I_n\, r}\right)^{\frac{1}{n+1}} \tilde{\sigma}_{ij}(\Theta, n) \ . \tag{2}$$

Here $\tilde{\sigma}_{ij}(\Theta,n)$ contains the dependence of stresses on the out-of-plane angle Θ, and r is the radial distance from the crack tip. I_n is an integration constant which depends on the creep exponent n.

Between the two limiting situations of K and C* control lies a transition period in which creep strains are higher than elastic strains over an extensive region ahead of the crack tip. During this period, the C* integral is path dependent. The time of this transition period (i.e., up to development of a path independent C* controlled HRR stress field) has been estimated approximately [7] as

$$t_I = \alpha_n \frac{K^2(1-\nu^2)}{(n+1)EC^*} \ , \tag{3}$$

where for plane-strain $\alpha_n = 0.96$, while for plane-stress $\alpha_n = 1.05$ (and the $(1-\nu^2)$ term is omitted). During the transition period, neither C* nor K are valid parameters describing the stress-field ahead of the crack. Crack propagation may well occur during the transition period, in which case, none of the steady-state parameters provides adequate correlation.

So far we have dealt only with analyses applicable to stationary cracks. The introduction of a moving crack greatly complicates the analysis. Even for room temperature fracture mechanics there is as yet, no complete description of the stress and strain fields for a stable moving crack. Hutchinson et al [10-12] and Rice et al [13] have shown that a milder stress-strain singularity then the HRR field, namely (1/ln(r)), may occur for creeping materials. Hui and Riedel [14] have given results for quasi-static crack growth under small-scale yielding conditions. They report a new type of singularity which depends on the creep exponent n such that (for n > 3),

$$\sigma_{ij} = \alpha_n \left[\frac{\dot{a}(t)}{B \, E \, r} \right]^{1/n-1} \tilde{\sigma}_{ij}(\Theta, n) \ ,$$

$$\dot{\varepsilon}_{ij} = \frac{\alpha_n}{E} \left[\frac{\dot{a}(t)}{B \, E \, r} \right]^{1/n-1} \dot{\varepsilon}_{ij}(\Theta, n) \ .$$

(4)

The amplitude of this singular field is related to the crack growth rate $\dot{a}$. This is a stronger singularity than that due to the HRR field. However, according to Parks [15] the stresses decay so rapidly with distance from the crack tip that this field may fail to dominate over a physically significant dimension.

In view of these and other uncertainties we have undertaken to analyse the time dependent variation of stress-strain fields ahead of a stationary crack using a Finite-Element Method. Our primary interest has been the approach to steady-state solutions. The results we obtain can be compared with analytical approximations for the transition period and are discussed in terms of their relevance to accelerated laboratory crack growth testing. The extension of this analysis to a moving crack is currently underway.

2. THE FINITE ELEMENT ANALYSIS

An incremental finite element program has been developed which follows the small-strain formulation of Kanchi et al [16]. For each time-step the stress increments $\Delta\sigma$, and strain increment $\Delta\varepsilon$, are calculated. An explicit time stepping procedure with variable time increment has been used (as suggested by Owen et al [17]). However, the cost effective explicit scheme is only conditionally stable, and the initial time increments were chosen according to the stability criterion proposed by Cormeau [18]. The incompressibility of creep deformation is utilized, by under integrating the element stiffness matrices [19]. Figure 1 shows the undeformed mesh containing 134 8-node isoparametric elements. The a/w ratio was kept constant at 0.5. Three geometries are studied using this mesh. For the double-edge notched (DEN) and centre-cracked panel (CCP) cases, the mesh represents one quarter of a specimen, while for the single-edge notched (SEN) case, half of a specimen is represented. Traction-free boundary conditions are applied to the crack surface, while nodes ahead of the crack in the crack plane are constrained to remain in that plane. Those nodes on the appropriate vertical symmetry axis are also constrained to model the DEN and CCP specimen geometries.

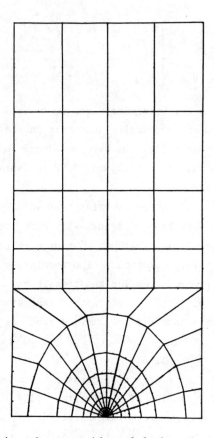

Figure 1 - The Finite Element Grid used during the Studies

The crack tip has been modelled using two different meshes. Initially, a small (r/a = 0.005) semicircular hole was placed at the tip. This was bounded by 12 equal sized elements, thus giving 25 nodes at the tip of the crack. In later work, this hole was removed, and the mesh refined from 134 elements to 204 elements. The crack tip was again modelled by 12 equal sized elements, but with one edge collapsed into a common point although free to deform. On the remaining edges, the midside nodes remained at their usual positions. This crack tip mesh configuration produces an HRR singularity for plastic solutions [21].

The material is assumed to obey a steady-state creep law described by equation (1). The parameters used are the same as those used by Riedel [8] and by Stonesifier et al [9], namely n = 5, B = 10^{-16} MN^{-5}/hr, E = 150GN/m^2 and ν = 0.3. Specimens are uniformly loaded from the ends to produce a stress intensity factor K of 40 MNm$^{-3/2}$ at the tip. At certain time intervals, the C(t) integral is calculated along several paths ahead of the crack tip, using

$$C(t) = \int \left[\left(\frac{n}{n+1} \sigma_{ij}\dot{\varepsilon}_{ij} - \sigma_{xx}\frac{\partial\dot{u}_{xx}}{\partial x} - \sigma_{yy}\frac{\partial\dot{u}_{yy}}{\partial x} \right) dy + \left(\sigma_{xy}\frac{\partial\dot{u}_{xx}}{\partial x} + \sigma_{yy}\frac{\partial\dot{u}_{yy}}{\partial x} \right) dx \right].$$ (5)

The integral paths chosen follow the edges of the elements. To do this, the stresses and strain rates calculated at the integration points are extroplated to the element nodal points [20]. The displacement rates appearing in equation (5) are approximated using,

$$\dot{u}_{ij} = \frac{\Delta u}{\Delta t}$$

where Δu is the displacement increment during the time increment Δt immediately prior to the calculation of C(t).

3. RESULTS

For all specimen geometries the initial elastic stress field ahead of the crack was kept constant by altering the magnitude of the applied loads. In each case, the elastic stress intensity factors calculated using an elastic finite element programme gave a maximum deviation of 0.96% from desired value of 40 MN m$^{-3\ell}$ and a maximum variation between the paths of less than 1 percent (using the grid given in Figure 1).

The variation of the C-integral with time is shown in Figures 2 - 4 for the SEN, CCP and DEN geometries respectively. C(t) values are calculated along each of 6 paths whose distance from the crack tip is P. The normalized path distance, d = P/c is also shown in the figures (c being the length of uncracked ligament). It is clear from these figures, that the progressive development of the creep zone ahead of the crack, causes convergence of C(t) values earliest for paths closest to the crack tip. Once C(t) values for all paths have converged to the same result, the calculations are continued for another 40 steps. This ensures complete convergence of C(t) to a steady-state C*. The average values obtained from 6 integral paths are 213 N/m hr for the SEN, 1017 N/m hr for the CCP and 417 N/m hr for the DEN geometry. While the SEN and CCP values agreed within 2 percent with the tabulated values of Shih and Kumar [22], for the DEN

geometry, the resulting value of C* is significantly higher than the
tabulated one.   For all geometries, the variation in C* between the paths
at steady-state was less than 5 percent of their average value.   The time
to reach a complete steady-state also varies considerably, from 12.4 hrs
for the CCP specimen to 49.4 for DEN and 116.8 hrs for SEN.

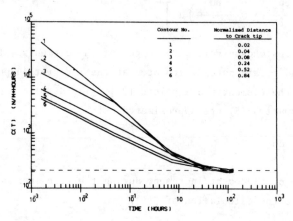

*Figure 2 - Variation of C(t) with Time and Integration Path for the SEN
Specimen Geometry (dashed line from Kumar and Shih [23])*

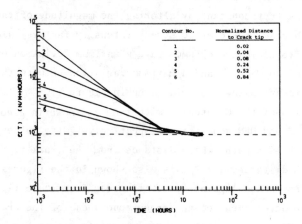

*Figure 3 - Variation of C(t) with Time and Integration Path for the CCP
Specimen Geometry (dashed line from Kumar and Shih [23])*

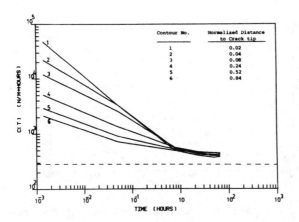

Figure 4 - Variation of C(t) with Time and Integration Path for the DEN Specimen Geometry (dashed line from Kumar and Shih [23])

The progressive development of the creep zones are given in Figures 5 and 6 for the SEN and CCP geometries. Here, the boundary is defined by the locus of points at which the equivalent creep strain is equal to the elastic strain ($\varepsilon_e = 2/3\ \varepsilon'_{ij}\ \varepsilon'_{ij})^{1/2}$). A linear interpolation of the plastic strains at the finite element integration points was used to construct the smooth boundaries. As the figures show, the creep zones have a similar shape to the time independent plastic zones found for small scale yielding conditions at lower temperatures. Also, until the creep zone extends over the whole area of the uncracked ligament, C* is path dependent.

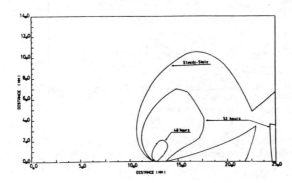

Figure 5 - Development of Creep Zones in the SEN Specimen

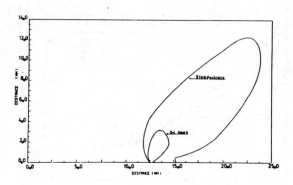

Figure 6 - Development of Creep Zones in the CCP Specimen

Figure 7 shows the accumulation of the effective creep strains at the first element ahead of the crack tip (0.125 mm). This is computed by averaging all points lying between 0 and 15 degrees from the crack plane. The steady-state displacement profiles of the cracks are given in Figure 8.

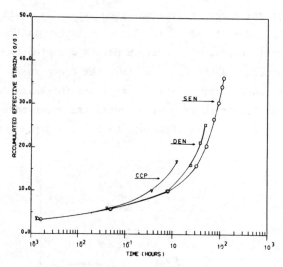

Figure 7 - Comparison of Accumulated Effective Creep Strains Ahead of the Crack Tip in the SEN, DEN and CCP Specimens

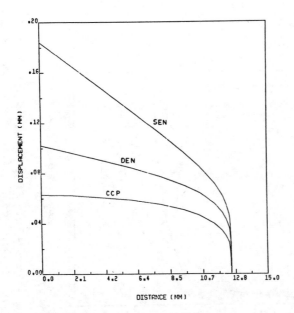

*Figure 8 - Crack Plane Displacement Profiles in the SEN, DEN and CCP
    Specimens at Steady-State*

The variation of the normalized tensile stress σ_{yy}/σ_0 with distance
along the ligament directly ahead of the crack is shown in Figures 9, 10
and 11. Here, the initial elastic and the analytical steady-state HRR
stress fields are also shown for comparison. Even after short time periods,
the stress field ahead of the crack tip deviates considerably from $1/\sqrt{r}$
singularity. At long times the steady-state stress-distribution for the
SEN specimen is in good agreement with the analytical HRR stress field up
to a distance of about 0.5 mm (or 5δ where δ is the crack tip opening).
For the DEN specimen agreement is fairly good but over a distance less
than that observed in SEN. For the CCP geometry there is no measurable
distance over which the steady-state stress distribution agrees with the
HRR field.

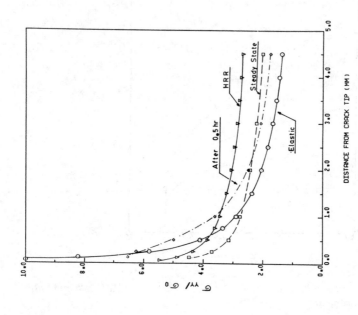

Figure 10 – Steady-State Stress Distribution Ahead of the Crack Tip (i.e., $\Theta = 0°$) in the DEN Specimen

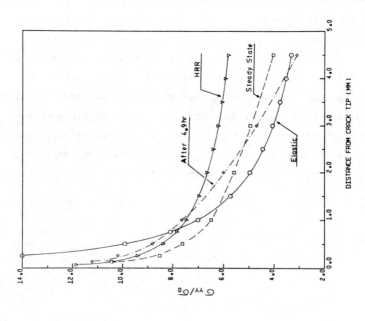

Figure 9 – Steady-State Stress Distribution Ahead of the Crack Tip (i.e., $\Theta = 0°$) in the SEN Specimen

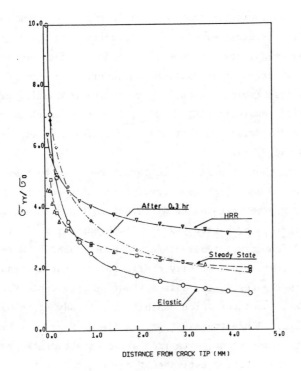

Figure 11 - Steady-State Stress Distribution Ahead of the Crack Tip
(i.e., Θ = 0°) in the CCP Specimen. Note "Δ" symbols are
from the solution which used a refined mesh (204 elements) and
collapsed 8 node isoparametric elements at the crack tip,
while the '□' symbols are due to the original solution which
uses a coarse mesh and a hole at the crack tip.

4. DISCUSSION

One would like to be able to characterize the state of stress and
strain surrounding a crack in a creeping solid, using a single parameter.
This could then be used with confidence to correlate experimental data
for creep crack growth and to help predict the behaviour of creeping structures which contain cracks. While this may be an achievable goal, there
will no doubt be restrictions on the use of any such parameter. These
must be defined before it can be used routinely and safely. In the present
work, we have attempted to study the effect of geometry on crack-tip
behaviour, in order to determine how this affects the development of steady-
state conditions, and the applicability of proposed crack tip parameters.

Experimental work indicates that of various possible crack-tip parameters, C* shows the greatest promise for characterizing crack growth behaviour. Our numerical results show that C* does indeed become path independent after a period of relaxation. However, even when identical initial elastic stress fields are applied to bodies containing cracks,[1] the time to reach a steady-state C* field t_{tr} and the amount of creep deformation required, varies substantially for different specimen geometrie (see Figures 2, 3, 4, 7, 8).

Let us deal with these two issues in turn. Because the time to reach steady-state varies so much with geometry, it is quite possible that, if one measures crack growth rates using different test specimens, one specimen may be characterized by a steady-state C* throughout most of the test, while for a second geometry, steady-state is never reached. Moreover, even if laboratory tests are carefully configured so that a steady-state is reached early in the tests, the data may not be applicable for estimating the life of a component which exhibits non-steady state behaviour for much of its life. Since transient stress fields near the crack tip are higher than at steady-state, data from steady-state crack growth experiments provide a nonconservative estimate of crack growth rate.

Regarding the absolute time for convergence, it is worth comparing our results with the estimate given by equation (3). This is derived by extrapolating an approximate solution for C(t) at short time, namely

$$C(t) = \alpha_n \frac{(1-\nu^2)K^2}{(n+1)Et}$$

to find the time t_I, at which it decreases to the steady-state C* value. While this is bound to underestimate the transition time, one might expect it to be reasonably close. However, the estimates of t_I for the present conditions are 1.5, 3.9 and 7.6 hours for the CCP, DEN and SEN specimens respectively. This compares with 12.4, 49.4 and 116.8 hr from finite-element calculations. Thus not only does equation 3 give estimates for t_{tr} which are short by over an order of magnitude, but the estimates become worse as C* decreases. More work is required to produce a reliable estimate for the transition period.

[1]
It is worth noting that because the elastic compliance of the CCP and DEN specimens are essentially the same, not only is K kept constant, but also the net section stress.

The strains accumulated during the transition period as shown in Figure 7 are substantial. It is therefore likely that cracks will start to propagate before a steady-state has been reached. The effect that this will have on the approach to steady-state is not clear. This is currently under investigation.

Figures 9, 10 and 11 compare the computed stress field with the analytical HRR fields. It is not clear that the distance ahead of the crack for which the HRR field is valid, is specimen dependent. For the CCP specimen no correlation was found (Figure 11). To investigate this further, the mesh was refined to 204 elements with collapsed elements at the tip (as discussed earlier). This has the effect of forcing an HRR singularity at the crack tip. However, the results obtained were no different from those given in Figure 11. Thus it would appear that for the conditions tested, the HRR field is applicable at the crack tip in a CCP specimen over a very small distance, and over a much smaller region than for the DEN and SEN specimens.

This raises the issue of validity criteria for C* testing. Such criteria have been suggested for J-integral tests [23]. It can be shown that in order to ensure an HRR field over a significant distance ahead of a crack tip, both specimen thickness and remaining ligament must exceed a critical value which depends on $(J/\sigma_{ys})^2$. No such criterion has been suggested for creep crack growth tests. However, something of this type is required if laboratory data are to be applied correctly to the analysis of creeping structures, in which a different geometry may drastically alter the size of the HRR field.

5. CONCLUSIONS

(1) When creeping structures containing cracks are placed under identical initial elastic loading conditions, the time to approach a steady-state C* stress field and the amount of creep deformation required is a sensitive function of the geometry.

(2) An approximate analytical expression (equation (3)) substantially underestimates the time required to reach steady-state.

(3) A set of validity criteria should be introduced for creep crack growth testing, which may be applied to both laboratory tests and the analysis of cracked components in service. These must ensure that steady-

state is reached early in the tests and that at steady-state an HRR field exists over a significant distance from the crack tip.

ACKNOWLEDGEMENTS

The authors wish to acknowledge funding for this work from the Natural Sciences and Engineering Research Council of Canada, and from the Department of Energy, Mines and Resources, through DSS Contract No. 05SU.23440-0-9199.

REFERENCES

1. SIVERNS, M.J. and PRICE, A.T., *Int. J. Fracture*, Vol. 2, 1973, p. 19
2. HARRISON, C.B. and SANDOR, G.N., *Eng. Fracture Mechanics*, Vol. 3, 1971, p. 403.
3. HAIGH, J.R., *Mat. Sci. Eng.*, Vol. 20, 1973, p. 213.
4. McEVILY, A.J. and WELLS, C.H., *Proc. of the Int. Conf. on Creep and Fatigue in Elevated Temperature Applications*, Philadelphia, Pennsylvania, 1973.
5. LANDES, J.D. and BEGLEY, J.A., *ASTM STP 590*, 1976, p. 128.
6. SAXENA, A., *ASTM STP 700*, 1980, p. 112.
7. RIEDEL, H. and RICE, J.R., *ASTM STP 700*, 1980, p. 112.
8. BASSANI, J.L. and McCLINTOCK, F.A., *Solids and Structures*, Vol. 17, 1981, p. 479.
9. STONESIFIER, R.B. and ATLURI, S.N., *Eng. Fracture Mechanics*, Vol. 16 1982, p. 625.
10. AMAZIGO, J.C. and HUTCHINSON, J.W., *J. Mech. Phys. Solids*, Vol. 25, 1977, p. 81.
11. DEAN, R.H. and HUTCHINSON, J.W., *ASTM STP 700*, 1980, p. 383.
12. HUTCHINSON, J.W. and PARIS, P.C., *ASTM STP 668*, 1979, p. 37.
13. RICE, J.R., DRUGAN, W.J. and SHAM, T.L., *ASTM STP 700*, 1980, p. 189.
14. HUI, C.Y. and RIEDEL, H., *Report MRS E-117*, Brown University, 1979.
15. PARKS, D.M., *Proc. of the Second Int. Conf. on Numerical Methods in Fracture Mechanics*, 1980, p. 239.
16. KANCHI, M.B., ZIENCIEWICZ, O.C. and OWEN, D.R.J., *Int. J. Numerical Methods in Fracture Mechanics*, Vol. 12, 1978, p. 109.
17. DINIS, L.M.S. and OWEN, D.R.J., *Computers and Structures*, Vol. 8, 1979, p. 207.
18. CORMEAU, I., *Int. J. Numerical Methods Eng.*, Vol. 9, 1975, p. 109.
19. NAYLOR, D.J., *Int. J. Numerical Methods Eng.*, Vol. 8, 1974, p. 443.
20. HINTON, E. and CAMPBELL, J.S., *Int. J. Numerical Methods Eng.*, Vol. 8, 1974, p. 461.
21. de LORENI, and SHIH, C.F., *Int. J. Fracture Mechanics*, Vol. 13, 1977, p. 507.
22. KUMAR, V. and SHIH, C.F., *EPRI Report No. 1931/1237-1*, 1981.
23. BEGLEY, J.A. and LANDES, J.D., *ASTM STP 514*, 1972, p. 1.

MODELLING PROBLEMS IN CRACK TIP MECHANICS   *Waterloo, Ontario, Canada*
CFC10, University of Waterloo   *August 24-26, 1983*

FATIGUE CRACK CLOSURE ANALYSIS AND CRACK PROPAGATION ESTIMATION

S.K.P. CHEUNG and I. LE MAY
Department of Mechanical Engineering
University of Saskatchewan
Saskatoon, Saskatchewan, Canada

1. INTRODUCTION

The purpose of this paper is to model analytically fatigue crack growth with a plasticity-induced crack closure effect in constant amplitude loading. The specimen considered in modelling is a centre-crack tension (CCT) one of 2219-T851 aluminium alloy.

In the model, the crack is considered to be in a continuum. Therefore, microstructural considerations involving grain size effects, grain interactions, inclusions, etc., which can have significant effects [1-3], are ignored. The short cracks studied are physically small as classified in [1].

It has been observed that short cracks grow much faster than do long cracks at equivalent values of stress intensity factor range, ΔK, based on linear elastic analysis. Short cracks also propagate at values of ΔK less than the threshold value for long crack growth, ΔK_T. Analysis of their propagation without considering closure effects is a risky procedure leading to nonconservative lifetime predictions. The fact that a short crack has small and limited compressive deformation along its wake behind the crack tip as compared with long cracks, so giving rise to reduced closure, is the major factor causing an increased propagation rate.

2. FATIGUE CRACK CLOSURE

Elber [4] noted that cracks closed at some positive load during constant cyclic loading. He postulated that cracks closed under the compressive residual plastic deformation left along the wake and generated at the crack tip. This was termed the "plasticity-induced crack closure" phenomenon.

The value of the stress which, upon unloading, causes the first element of the crack surface just to close is termed the contact or closure stress. Upon reloading, the applied stress just to fully open the crack is the opening stress (σ_{op}).

Thus, the Paris law should incorporate an effective stress range factor, U, in its equation to accommodate the closure effect, and we may rewrite the crack growth law in the form:

$$da/dN = C(U\Delta K)^m F , \qquad (1)$$

where C and m are material constants and F is the boundary-correction factor.

$$U = (\sigma_{max} - \sigma_{op})/(\sigma_{max} - \sigma_{min}) , \qquad (2)$$

where σ_{max} and σ_{min} are respectively the maximum and minimum applied stress in constant amplitude loading. The effective stress range factor depends on the stress ratio R $(= \sigma_{min}/\sigma_{max})$, the ratio of applied stress to yield stress, and the cyclic strain hardening exponent, n, of the material [5-7]. Since U depends only on opening stress at the particular applied loading level in constant amplitude loading, opening-stress will also have the same dependent factors as the effective stress range factor U.

The analytical closure model utilized here is based on a crack growth equation which accounts for growth rates approaching critical and threshold stress intensity factors [8,9]. It has the form:

$$da/dN = C_1(\Delta K_{eff})^{C_2}[1-(\Delta K_o/\Delta K_{eff})^2]/[1-(K_{max}/C_5)^2] , \qquad (3)$$

$$\Delta K_o = C_3[1-C_4(\sigma_{op}/\sigma_{max})] , \qquad (4)$$

$$K_{max} = \sigma_{max}\sqrt{\pi a}\, F , \qquad (5)$$

and

$$\Delta K_{eff} = (\sigma_{max} - \sigma_{op})\sqrt{\pi a}\, F , \qquad (6)$$

where C_1, C_2, C_3, C_4, C_5 are all constants based on experimental data. The data used in the model were taken from references [8,9].

It may be noted here that crack closure has been directly observed experimentally in plane stress conditions by a number of investigators [4,10-12].

2.1 Crack Closure Model

There have been a number of attempts to analyze crack closure. El Haddad and co-workers [13,14] suggested an effective crack length method and a J-integral method in explaining faster short crack growth. Some investigator have used the finite element method [15-18] and this has proved to give good, reliable results, but uses extensive computer time. In the present case, Newman's analytical crack closure modelling is used, the details being given in references [8,9,19].

The model studied utilized a finite width (W = 0.1524 m) CCT specimen. The crack surface displacements were obtained for an infinite CCT plane from [20] using Westergaard stress functions, and the opening stresses, σ_{op}, were calculated using elastic and elastic-plastic fracture mechanics analysis. The crack opening displacement equations are based on a modified Dugdale model loaded with uniform remote stress and uniform stress applied over the crack tip segment of crack surface. One quarter of the plate with a CCT crack is shown in Figure 1 with the coordinate system specified.

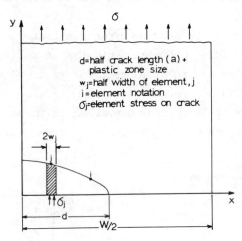

Figure 1 - The CCT Crack under Uniform Loading

The material considered was 2219-T851 aluminium alloy with the following properties: yield stress, σ_y = 360 MPa; UTS, σ_u = 455 MPa; flow stress, σ_f = 407.5 MPa; elastic modulus, E = 73 GPa; Poisson's ratio, ν = 0.25.

The material constants for the two constraint factors (α) are as follows [8,9].

$\alpha = 1.9$	$\alpha = 2.3$
$C_1 = 2.486 \times 10^{-10}$	$C_1 = 1.764 \times 10^{-10}$
$C_2 = 3.115$	$C_2 = 3.18$
$C_3 = 2.97$	$C_3 = 2.97$
$C_4 = 0.8$	$C_4 = 0.8$
$C_5 = 77$	$C_5 = 77$

3. RESULTS AND DISCUSSION

In the model, the crack propagation rate depends on the crack length and the load history. The model may be used to calculate the growth of both short and long cracks.

The residual stress is built up after many cycles of alternating plastic flow displacements at the tip. Upon unloading to minimum load, the normalized contact stresses, σ_j/σ_f, along the crack are shown in Figure 2. These computed normalized contact stresses are similar to those in [21] considering an edge dislocation in approximating the maximum principal residual stress component, and are also very close to Newman's results [8].

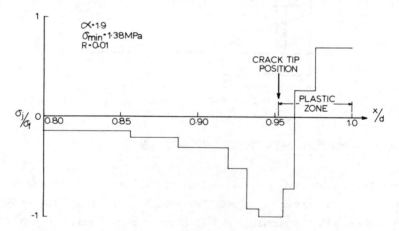

Figure 2 - Computed Contact Stresses Along the Crack at Minimum Stress with Constant Amplitude Loading

The compressive force created from contact stresses is the closing force produced by plastic deformation. When the crack is short, the compressive force is small because the contact area is limited. Since there is not sufficient compressive force on the short crack surfaces to close the crack as compared to the situation for the long crack [22], the short crack grows faster. The total compressive force increases as the crack grows.

The smaller the compressive force, the easier will the crack be opened. The growth rate is high during initial growth with negligible prior deformation. The compressive force at the tip contributes as much compression as does that left in the wake during propagation of a short crack. However, the compressive force produced from the residual deformation left in the wake becomes much greater than that from the tip as the crack grows, particularly when the loading is under positive R values. The crack tip may be under tensile force when the crack is subjected to plane strain loading with positive R. From the computed results, negative values of R give rise to an increasingly compressive force at the crack tip plastic zone, while with high positive values of R a tensile residual force may be produced at the tip during its growth. The crack is under more compressive force if plane stress loading is applied rather than plane strain.

There is more compression if the crack is under more negative R loading, while the total force on the crack may be tensile with positive R loading. In other words, the crack may not close during cycling and the closure effect for a short crack with R greater than zero is negligible.

In Figure 3, the large drop in growth rate during early crack propagation for $R = -1$ is due to the significant closure in the wake of the advancing crack. The positive R ratios sometimes barely produce the short crack effect; this happens particularly when the crack is under plane strain conditions. The small drop in crack growth rate during early crack growth with positive R is due to the initial build up of the compressive residual deformation along the wake.

After some crack extension, there will be more contact area available. Thus, the growth rate decreases and higher opening stresses are required. The growth rate starts to increase when the crack extension effect just cancels the closure effect. The turning point, a_T, is the crack length where a short crack meets the long crack criterion as may be seen in Figure 4. Positive R values produce a smaller a_T than do negative R values.

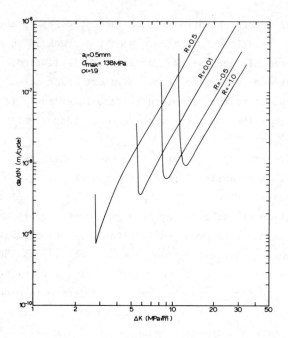

Figure 3 - Crack Growth Rate Plots as a Function of the R Ratio for
$\sigma_{max} = 138$ *MPa*

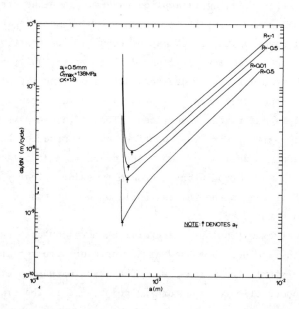

Figure 4 - Crack Growth Rate Curves of Figure 3 Plotted as a Function of
Crack Length

As noted previously the closure effect is more prominent in plane
stress conditions than under plane strain, and this is illustrated in
Figure 5. This is mainly due to the plastic zone size around the crack
tip being smaller in plane strain loading, producing reduced crack closure.
Cyclic loading under plane stress conditions produces a higher opening
stress, which means cracks may close where the plane stress state occurs,
e.g., at the surface, before they do under the plane strain state, as
normally found away from the surface.

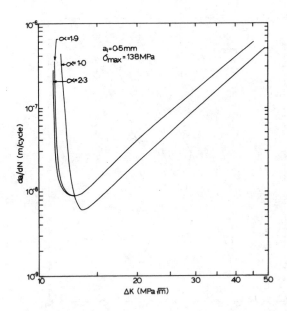

Figure 5 - Crack Growth Rate Curves as Affected by the Degree of Constraint

In the analysis, the plane stress or plane strain condition is assumed
throughout the crack growth, i.e., α is assumed to be constant. Improvement in crack prediction could be expected if the transition from plane
stress to plane strain during growth were taken into account. As discussed
in [11], the constraint factor is not constant throughout the plate thickness nor during the propagation, but the assumption of constant α is utilized
here to simplify the problem.

The smaller the value of σ_{max}, the larger is the crack closure when
loading for a fixed value of R. This can be seen by comparing Figures 3
and 6. The smaller the maximum stress, the higher the normalized opening

stress (σ_{op}/σ_{max}), and the greater is the closure effect. To maximize
the closure effect, the crack should be under small maximum applied stress,
plane stress conditions and large negative R. In other words, large σ_{max},
plane strain and high positive R loading are all factors that give a crack
almost zero closure effect. At negative R (e.g., $R = -1$) loading, the
crack will propagate most readily at large values of σ_{max} and under plane
strain conditions.

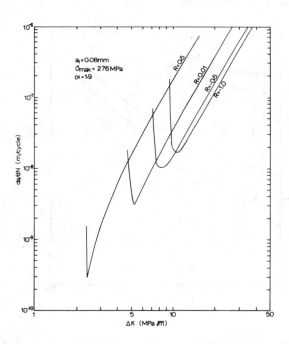

Figure 6 - Crack Growth Rate Plots as a Function of the R Ratio for
$\sigma_{max} = 276$ *MPa*

It may be seen that using Newman's model for crack closure the observed
form of the growth behaviour of short cracks is predicted. In the model of
Lal and Le May [6], for example, the closure force is treated as being
ahead of the crack tip and the crack length does not enter the picture:
the Lal and Le May model cannot, therefore, be used for short cracks.

Further steps to be taken in the present study include the comparison of
predicted and experimental growth data for short and long cracks, and evalua‑
tion of the effects of material properties and of variable amplitude loading
on growth.

REFERENCES

1. RITCHIE, R.O. and SURESH, S., "Mechanics and Physics of the Growth of Small Cracks", *Proc. of the 55th AGARD Meeting on Behavior of Short Cracks in Airframe Components*, Toronto, 1982.
2. MORRIS, W.L., "The Noncontinuum Crack Tip Deformation Behavior of Surface Microcracks", *Met. Trans. A*, Vol. 11A, 1980, pp. 1117-1123.
3. MORRIS, W.L., JAMES, M.R. and BUCK, O., "Growth Rate Models for Short Surface Cracks in Al 2219-T851", *Met. Trans. A*, Vol. 12A, 1981, pp. 57-64.
4. ELBER, W., "Fatigue Crack Closure Under Cyclic Tension Loading", *Eng. Fracture Mechanics*, Vol. 2, 1970, pp. 37-45.
5. LAL, K.M., GARG, S.B.L. and LE MAY, I., "On the Effective Stress Range Factor in Fatigue", *J. Eng. Mater. and Technology*, Vol. 102, 1980, pp. 147-152.
6. LAL, K.M. and LE MAY, I., "An Assessment of Crack Closure in Fatigue Using the Westergaard Stress Function", *Fatigue of Engineering Materials and Structures*, Vol. 3, 1982, pp. 99-111.
7. LE MAY, I., "Fatigue Damage Mechanisms and Short Crack Growth", *Proc. of the 55th AGARD Meeting on Behavior of Short Cracks in Airframe Components*, Toronto, 1982.
8. NEWMAN, J.C., Jr., "A Crack-Closure Model for Predicting Fatigue Crack Growth Under Aircraft Spectrum Loading", *ASTM STP 748*, 1981, pp. 53-83.
9. NEWMAN, J.C., Jr., "Prediction of Fatigue Crack Growth Under Variable Amplitude and Spectrum Loading Using a Closure Model", *ASTM STP 761*, 1982, pp. 255-277.
10. CHENG, Y.F., and BRUNNER, H., "Photoelastic Research in Progress on Fatigue Crack Closure", *Int. J. Fracture Mechanics*, Vol. 6, 1970, pp. 431-434.
11. LINDLEY, T.C. and RICHARDS, C.E., "The Relevance of Crack Closure to Fatigue Crack Propagation", *Materials Sci. and Engineering*, Vol. 14, 1974, pp. 281-293.
12. SHAW, W.J.D. and LE MAY, I., "Crack Closure During Fatigue Crack Propagation in Fracture Mechanics", *ASTM STP 677*, 1979, pp. 233-246.
13. TOPPER, T.H. and EL HADDAD, M.H., "Fatigue Mechanics Analysis for Short Fatigue Cracks", *Proc. V. Inter-American Conference on Materials Technology*, São Paulo, Brazil, 1978, pp. 493-500.
14. EL HADDAD, M.H., SMITH, K.N. and TOPPER, T.H., "Fatigue Crack Propagation of Short Cracks", *J. Engineering Materials and Technology*, Vol. 101, 1979, pp. 42-46.
15. NEWMAN, J.C., Jr., "Finite-Element Analysis of Fatigue Crack Propagation - Including the Effects of Crack Closure", *Ph. D. Thesis*, Virginia Polytechnic Institute and State University, Blacksburg, Virginia, May, 1974.
16. OHJI, K., OGURA, K. and OHKUBO, Y., "Cyclic Analysis of a Propagating Crack and Its Correlation with Fatigue Crack Growth", *Eng. Fracture Mechanics*, Vol. 7, 1975, pp. 457-464.
17. NEWMAN, J.C., Jr., "A Finite-Element Analysis of Fatigue Crack Closure", *ASTM STP 590*, 1976, pp. 281-301.
18. NEWMAN, J.C., Jr., "Finite-Element Analysis of Crack Growth Under Monotonic and Cyclic Loading", *ASTM STP 637*, 1978, pp. 56-80.
19. NEWMAN, J.C., Jr., "A Nonlinear Fracture Mechanics Approach to the Growth of Small Cracks", *Proc. of the 55th AGARD Meeting on Behaviour of Short Cracks in Airframe Components*, Toronto, 1982.

20. TADA, H., PARIS, P.C. and IRWIN, G.R., *The Stress Analysis of Cracks Handbook*, Del Research Corporation, 1973.
21. TIROSH, J. and LADELSKI, A., "Note on Residual Stresses Induced by Fatigue Cracking", *Eng. Fracture Mechanics*, Vol. 13, 1980, pp. 453-46
22. COOK, R., EDWARDS, P.R. and ANSTEE, R.F.W., "Crack Propagation at Short Crack Lengths under Variable Amplitude Loading", *Proc. of the 11th ICAF Symposium*, Noordwijkerhout, The Netherlands, May, 1981. (ICAF Doc. No. 1216), pp. 2.8/1 - 2.8/29.

MODELLING PROBLEMS IN CRACK TIP MECHANICS   *Waterloo, Ontario, Canada*
CFC10, University of Waterloo   *August 24-26, 1983*

DETERMINING STRESS INTENSITY FACTORS FOR RUNNING CRACKS

R. CHONA, W.L. FOURNEY and R.J. SANFORD   A. SHUKLA
University of Maryland   University of Rhode Island
College Park, Maryland, U.S.A.   Kingston, Rhode Island, U.S.A.

1. INTRODUCTION

Analyses of running cracks in large structures have been performed using dynamic finite-element computer codes [1,2,3]. When predicting crack extension behaviour, these codes generally require that the relationship between crack velocity, c, and instantaneous stress intensity factor, K_I, be specified. Dynamic photoelasticity, in both transmission and reflection modes, has proved to be a useful tool for determining this relationship experimentally [4]. However, in most cases, extraction of dynamic K-values from experimental mechanics data has relied on a one or two parameter representation of the stress field [4,5,6]. This makes it difficult to be confident of the accuracy of K-determination and is particularly critical if the results obtained experimentally form the basis for further calculations.

The stress field representation needed to completely describe the stress state around static cracks in fracture specimens of finite dimensions has been studied extensively [7,8,9]. This paper develops the counterpart stress field equations for plane elastodynamic crack problems which, when combined with the multiple-point, least-squares method [10,11], allow dynamic values of K and additional stress field parameters to be evaluated from photoelastic data with a high degree of accuracy.

2. DYNAMIC STRESS FIELD REPRESENTATION

Irwin [12] has shown, that for a crack tip stress field translating in the positive x-direction at a fixed speed, the dilatation, Δ, and rotation, ω, can be expressed as

$$\Delta = \frac{\partial u}{\partial x} + \frac{\partial v}{\partial y} = A(1-\lambda_1^2) \ \mathrm{Re} \ \Gamma_1(z_1) \ ,$$

$$\omega = \frac{\partial v}{\partial x} - \frac{\partial u}{\partial y} = B(1-\lambda_2^2) \ \mathrm{Im} \ \Gamma_2(z_2) \ ,$$

$$(1)$$

where λ_1, λ_2, z_1 and z_2 are defined in Figure 1. Using Hooke's Law

$$\sigma_{xx} = \mu[A(1+2\lambda_1^2-\lambda_2^2) \text{ Re } \Gamma_1 - 2B \lambda_2 \text{ Re } \Gamma_2] \ ,$$

$$\sigma_{yy} = \mu[-A(1+\lambda_2^2) \text{ Re } \Gamma_1 + 2B \lambda_2 \text{ Re } \Gamma_2] \ , \tag{2}$$

$$\tau_{xy} = \mu[-2A \lambda_1 \text{ Im } \Gamma_1 + B (1+\lambda_2^2) \text{ Im } \Gamma_2] \ ,$$

where the constants A and B have to be evaluated so as to satisfy the boundary conditions for the crack problem of interest and depend upon the choice of the stress functions, Γ_1 and Γ_2. For the opening mode case, on logical choice is

$$\Gamma_1 = \sum_{n=o}^{n=N} C_n \, z_1^{n-1/2} \quad \text{with} \quad \Gamma_2 = \sum_{n=o}^{n=N} C_n \, z_2^{n-1/2} \ . \tag{3}$$

The leading coefficient, C_0, is related to the stress intensity factor, $K = C_0\sqrt{2\pi}$, and the leading term is the familiar inverse-square-root stress singularity. The symmetry condition requires that $B = 2\lambda_1 A/(1+\lambda_2^2)$.

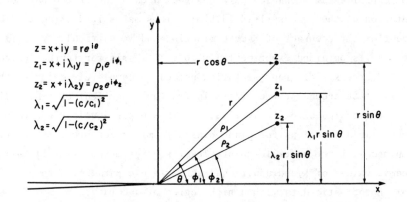

Figure 1 - The Coordinate Systems and Transformation Relations Used in the Constant-Crack-Speed Analysis

For the static problem, it has been demonstrated [2,8] that a second stress function is required to completely describe the stress field in specimens with finite boundaries. The analogous choice for the dynamic problem is

$$\Gamma_1 = \sum_{m=o}^{m=M} D_m \, z_1^m \quad \text{with} \quad \Gamma_2 = \sum_{m=o}^{m=M} D_m \, z_2^m \ . \tag{4}$$

In this case, the boundary condition on the crack faces requires that $B = A(1+\lambda_2^2)/2\lambda_2$. The leading term, D_0, gives rise to a superposed constant stress in the direction of crack propagation which is similar to the σ_{ox}-term in Irwin's static near-field equations [13]. This term can also be regarded as the far-field biaxiality correction factor studied extensively by Liebowitz and his co-workers [7] for the static case.

3. PARAMETER DETERMINATION USING THE LEAST-SQUARES METHOD

Equations (2) - (4) when combined with the stress-optic law can be used to relate the fringe order, N, at any point in an isochromatic field with the unknown real coefficients, C_n and D_m,

$$(Nf_\sigma/2t)^2 = \tau_{max}^2 = \frac{1}{4}(\sigma_{yy}-\sigma_{xx})^2 + \tau_{xy}^2 , \tag{5}$$

where f_σ is the fringe sensitivity of the model material and t is the model thickness. The first step in the analysis of an isochromatic pattern is to take a region around the crack tip from the experimental pattern being analyzed, extract a large number of individual data points and determine the coordinates and fringe order at each point. These data points are then used as inputs to an over-determined system of nonlinear equations of the form of equation (5) and solved in a least-squares sense for the unknown coefficients. As a final check, the best-fit set of coefficients is used to reconstruct the fringe pattern over the region of data acquisition to ensure that the computed solution set does, in fact, predict the same stress distribution as that observed experimentally.

When analyzing dynamic stress patterns, the data acquisition region is usually restricted to that portion of the stress pattern that can be seen to translate with only moderate changes in order to approximate the constant crack-speed assumption. In cases where the crack tip is approaching a specimen boundary, the data region should be restricted to no more than 1/4 to 1/3 the distance to the boundary. The number of coefficients necessary for an adequate representation of the stress field over the data acquisition region can be estimated by examining, as a function of the number of parameters, the average fringe order error, the values of the leading coefficients, and the reconstructed (computer-generated) fringe pattern corresponding to a given set of coefficients. Stability of the leading coefficients, as well as a good visual match between experimental and reconstructed patterns indicates convergence to a satisfactory solution

4. ILLUSTRATIVE EXAMPLE

The methodology outlined above is illustrated through the analysis of a crack propagation experiment performed using a Homalite 100 Ring segment whose geometry and loading are shown in Figure 2. A total of 16 flash photographs of the isochromatic fringe patterns associated with the running crack over the range $a/W = 0.19 - 0.90$ were recorded using a Cranz-Schardin type high-speed camera system [4]. Figure 3 shows the isochromatic pattern 145 μs after crack initiation, when the crack tip was located at an a/W of 0.52, and the crack speed was 15,000 inches/sec (375 m/s; $c/c_2 = 0.31$). Also shown in Figure 3 is a circular region of radius $0.15W$ from which a total of 60 data points were taken for analysis purposes. This data set was input to the least-squares algorithm and analyzed using successively higher order models. Figure 4 shows the changes in the error term and in the coefficients, C_0, D_0 and C_1, that occurred as the order of the analysis model was increased from two parameters (C_0,D_0) to six parameters $(C_0,D_0,C_1,D_1,C_2,D_2)$. Figure 5 compares the experimental fringe pattern over the data acquisition region with the computer-generated patterns corresponding to the coefficient set from each model. It can be seen that six parameters are required before the reconstructed pattern matches the salient features of the input experimental pattern. The same conclusion could be reached from an examination of the error term in Figure 4, where the stabilized error of 2.5% associated with a six-parameter model has been found to be typical of the error at which a good match is achieved.

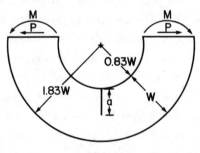

RING SPECIMEN

Figure 2 - The Geometry and Loading of the Ring Segment Used

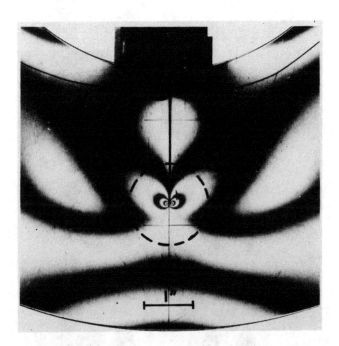

Figure 3 - The Isochromatic Fringe Pattern 145 μs after Crack Initiation in the Ring Segment; a/W = 0.52; c/c₂ = 0.31

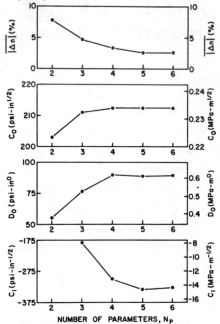

Figure 4 - The Error Term and Leading Coefficients versus the Number of Parameters in the Analysis Model

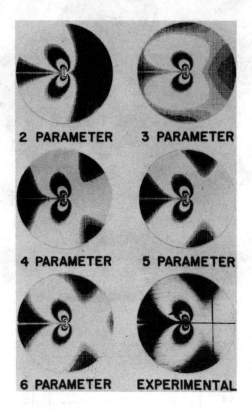

Figure 5 - Experimental and Reconstructed Fringe Patterns from Different Order Models for the Fringe Pattern of Figure 3

A similar analysis procedure was followed for all sixteen frames, and Figure 6 shows the instantaneous crack-tip stress intensity factor as a function of the crack tip position for this experiment. The decreasi K-field is typical of this particular geometry and loading, and the result shown were obtained from analyses with parameter models ranging from four to seven, depending upon the crack tip position and the size of the data region relative to the specimen boundaries. The variation with a/W of the dimensionless values of the two leading nonsingular terms, D_0 and C_1, is shown in Figure 7. These parameters are seen to behave in a systematic fashion as the crack length increases and their general trends are similar to previous results from other specimen geometries [9]. The influence of the approaching specimen boundary can be clearly seen in the rapid changes of both parameters beyond an a/W of 0.8.

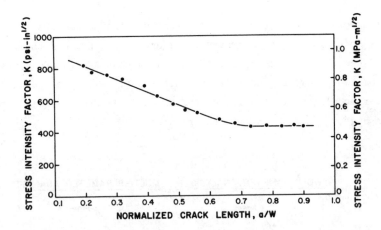

Figure 6 - *Instantaneous Stress Intensity Factor as a Function of Crack Length for a Crack Propagating Across a Ring Segment*

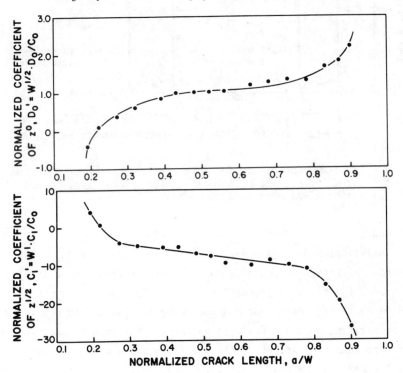

Figure 7 - *The Variations with Crack Length of the Two Lowest-Order Nonsingular Terms for a Crack Propagating Across a Ring Segment*

The importance of including more than just two parameters in the analysis model is best illustrated by Figure 8, which shows the relationship between the crack speed and the instantaneous K-value for this experiment from two-parameter and multi-parameter analyses. The use of higher parameter models substantially reduces the data scatter and leads to a greater degree of confidence in the results.

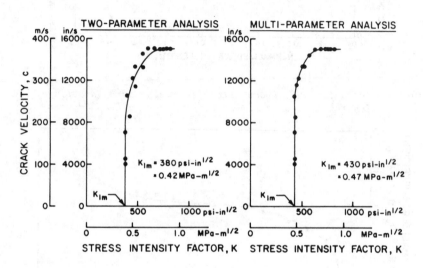

Figure 8 - The c-K Relationship for Homalite 100 Based upon Two-Parameter and Multi-Parameter Analyses of the Dynamic Isochromatic Fringe Patterns Recorded in a Ring Segment

5. SUMMARY

An analytical model and a parameter extraction methodology have been presented, which together allow an accurate representation of the stress state around the tip of a propagating crack. An example has been presented in which this technique has been used to analyze the behaviour of a crack propagating across a fracture specimen. The results show that the use of more than two parameters in the stress field representation substantially reduces the scatter in the experimentally determined K-values. The lower-order nonsingular terms have been shown to vary systematically with crack tip position and the role that these additional parameters play in the fracture process can now be studied.

ACKNOWLEDGEMENTS

The research reported here has been supported by the United States Nuclear Regulatory Commission and Oak Ridge National Laboratory at the University of Maryland, and by the United States National Science Foundation at the University of Rhode Island.

REFERENCES

1. HAHN, G.T., et al, *Report BMI-1937*, Battelle, 1975.
2. IRWIN, G.R., et al, *Report NUREG/CR-1455*, U.S. Nuclear Regulatory Commission, 1980.
3. NISHIOKA, T. and ATLURI, S.N., *Eng. Fracture Mechanics*, Vol. 16, 1982, pp. 157-175.
4. DALLY, J.W., *Experimental Mechanics*, Vol. 19, 1979, pp. 349-361.
5. BEINERT, J. and KALTHOFF, J.F., *Mechanics of Fracture VII*, 1982.
6. RAMULU, M. and KOBAYASHI, A.S., *Experimental Mechanics*, Vol. 23, 1983, pp. 1-9.
7. LIEBOWITZ, H., et al, *Engineering Fracture Mechanics*, Vol. 10, 1978, pp. 315-335.
8. SANFORD, R.J., *Mechanics Research Communications*, Vol. 6, 1979, pp. 289-294.
9. CHONA, R., IRWIN, G.R. and SANFORD, R.J., *ASTM STP 791*, 1983, pp. I-3 - I-23.
10. SANFORD, R.J. and DALLY, J.W., *Engineering Fracture Mechanics*, Vol. 11, 1979, pp. 621-633.
11. SANFORD, R.J., *Experimental Mechanics*, Vol. 20, 1980, pp. 192-197.
12. IRWIN, G.R., *Lehigh University Lecture Notes*, 1968.
13. IRWIN, G.R., *Proc. SESA, XVI*, 1958, pp. 93-96.

MODELLING PROBLEMS IN CRACK TIP MECHANICS
CFC10, University of Waterloo

Waterloo, Ontario, Canada
August 24-26 1983

EXPERIMENTAL MECHANICS AND MODELLING OF FATIGUE CRACK GROWTH

David L. DAVIDSON
Southwest Research Institute
San Antonio, Texas, U.S.A.

1. INTRODUCTION

The dynamic observation of crack growth under cyclic loading conditions by the use of high resolution electron optical techniques is a logical step in the continuing study of fatigue failure mechanisms. Examination of fatigue fracture surfaces with electron microscopes has been standard practice for many years, and led, early in its use, to the discovery of periodic markings on the fracture surface called striations. Several years ago we made the decision to couple a cyclic loading machine with the scanning electron microscope in order to utilize the resolution and depth of field of the SEM for observing fatigue crack growth dynamically [1]. Detailed examination of photographs made in the SEM at the extremes of the loading cycle led to the discovery of the stereoimaging technique for visualizing displacements [2] and computing strains [3]. Since the development of these new tools, fatigue crack growth in a variety of alloys has been investigated at very high resolution, as compared to methods previously used.

The only variables which have been systematically studied so far are the effects of cyclic stress intensity factor (ΔK) and environment (water vapor versus vacuum). The effects of metallurgical parameters [4], temperature [5], crack size [6] and nonuniform loading amplitude have been investigated, but not systematically. The author has also developed a semi-theoretical model for bringing together metallurgical factors and measured crack tip parameters. The purpose of this brief paper is to present some of the key experimental findings which relate directly to fatigue crack growth, together with a summary of the model for crack growth which uses these results.

2. DYNAMIC OBSERVATIONS OF FATIGUE CRACK GROWTH

The cyclically loaded fatigue crack may be viewed at the actual rate of loading within the SEM by using television scan rates. This system also allows crack growth to be videotaped for further study [7]. Another method of obtaining useful information is to periodically record the crack tip at maximum load on either videotape or by still photography, so that the change in crack tip shape and position may be directly correlated with cycles and microstructure. Such a sequence is shown in Figure 1, which illustrates *the important finding that cracks do not grow on each cycle;* rather crack growth only occurs periodically, with the number of cycles between growth increments directly dependent on the magnitude of the drivi force ΔK. For the case illustrated, there are a relative large number of cycles expended prior to crack growth, followed by a relatively small numb of cycles during which the crack extends. The events shown in this figure are repeated irregularly, so that it is difficult to determine experimentally the average number of cycles which must be applied in preparation for an increment of crack advance, but the process clearly consists of two steps: (1) cyclic damage accumulation and (2) crack advance. This two-step process leaves periodic marks on the fracture surface - striations. The detailed examination of striations by Nix and Flower [8] using scannir and transmission electron microscopy confirms that striation formation is a two-step process. This result correlates very well with our dynamic observations.

The sequence illustrated in Figure 1 has also been observed in two pow der aluminium alloys made by different metallurgical processes and an austenitic stainless steel. The process of crack advance is a little different in each material but is generally the same as shown in the figure. With increasing ΔK, the number of loading cycles prior to crack advance decreases until at some value, the crack advances on each cycle. A decrea in the number of cycles prior to crack advance with increasing ΔK has been experimentally verified, although not well quantified. During the period of cycling prior to crack advance, the crack tip region does not remain the same. Microstructure appears to control both the way the crack tip deforms during the cycles leading to crack advance, as well as the amount of crack growth which occurs. The next section discusses how this is determined and what, in fact, occurs.

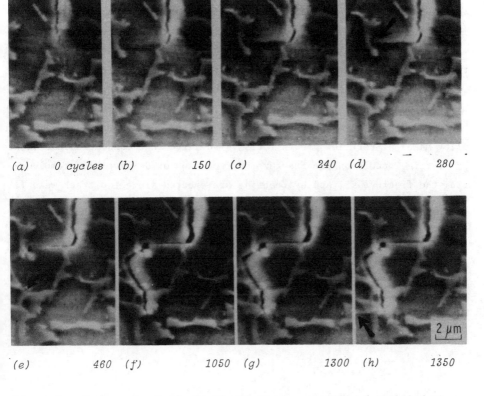

(a)   *0 cycles*   *(b)*   *150*   *(c)*   *240*   *(d)*   *280*

(e)   *460*   *(f)*   *1050*   *(g)*   *1300*   *(h)*   *1350*

Figure 1 - Fatigue Crack Tip in Ti-6A1-4V, Recrystallized Annealed Micro-
structure Showing Discontinuous Crack Growth. Crack does not
grow for the first 240 cycles shown (a-c), but a slip line
does form. Between 240 and 280 cycles, the crack grows rapidly
(arrow). The crack continues to grow rapidly from (d) to (e)
over the next 220 cycles (arrow). From (e) to (f), 600 cycles
are required for the crack to advance 1.5 μm. Between (f) and
(h), 350 cycles, the crack does not advance, but another slip
line is formed (arrow). $\Delta K = 18\ MN/m^{3/2}$

3. MEASUREMENT OF CRACK TIP PARAMETERS

The stereoimaging technique [2] allows displacements in the plane of
the photograph to be visualized and measured; this is illustrated in
Figure 2 for a Mode I, pin-loaded, single-edge-notched specimen. The
nonsymmetry of the displacements about the line of the crack is quite
typical, and is considered to be an important result of these measurements.
The uniform displacement state which exists in the net section of the
specimen is clearly changed to a biaxial state in the vicinity of the
crack tip. Crack opening is also a mix of Mode I and Mode II displacements,

where the magnitude of Mode II has been found to depend on microstructure. This is illustrated in Figure 3, which compares the crack opening displacements (COD) of two aluminium alloys [9,10]. So far, it has not been possib to determine the characteristics of microstructures in which a large fraction of crack opening displacement is Mode II; however, for the aluminium alloys studied, a decreasing ΔK results in an increasing fraction of COD being Mode II. The magnitude of Mode II for the powder metallurgy aluminiu alloy MA-87 at very low growth rates is so large that considerable evidence of fretting occurs on the fracture surface, and oxide particles removed from the fracture surface by fretting are ejected along the crack wake [11]

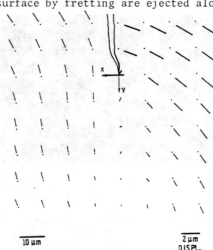

$\overline{10 \, \mu m}$ $\overline{2 \, \mu m}$
                                                         DISPL.

Figure 2 - Displacements Near a Fatigue Crack Tip in the Aluminium Alloy IN-9052. Loading was in the x-direction, $\Delta K = 9.6 \ MN/m^{3/2}$. Original location of the point is at the dot; displacement is to the end of the line. Note displacements are shown enlarged.

Fatigue cracks have been found to be tightly closed at the minimum in the load cycle for all the alloys studied so far. Stress ratios (R) up to 0.6 have not resulted in the crack being open at minimum load. Rapid changes in R, underloads and overloads, however, result in cracks which do remain open at minimum load.

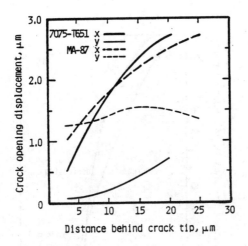

Figure 3 - COD Parallel and Perpendicular to Loading Direction for Two Aluminum Alloys at $\Delta K = 8$ MN/m$^{3/2}$ in Vacuum. MA-87 shows a large Mode II (y-direction) component.

Upon loading a crack which is tightly closed at minimum load, the crack begins to open at the crack mouth, but not at the crack tip. As load is increased the crack continues to "peel" open until it is fully open down to the tip. This process is nonlinear, as is shown in Figure 4. The opening load is dependent on ΔK, increasing as ΔK is decreased. At $\Delta K \sim K_{Ic}/2$ for 7075-T651, for which the most measurements have been made, $K_{op}/K_{max} \cong 0.6$, which is typical of values measured for other alloys. The origin of this so-called "crack closure" effect is not clear, and it will require considerably more effort to determine whether or not it is due to plasticity along the crack flank or residual stress ahead of the crack tip.

From the measured displacements, e.g., Figure 2, strains may be computed. Three elements of the symmetric strain tensor ε_{xx}, ε_{yy} and γ_{xy} may be determined, and through a Mohr's circle construction, the principal strains ε_1 and ε_2 and the maximum shear strain γ_{max} may also be computed. These are total strains, elastic plus plastic, and are actually strain increments because they result from an incremental load change, and are referenced to the minimum load condition. Both small strain and large strain theory have been used for this computation, with the latter yielding about a ten percent increase over the former. Large strain theory is used routinely.

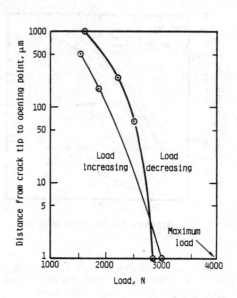

Figure 4 - Opening Sequence for Fatigue Crack in Ti-6A1-4V Titanium Alloy at $\Delta K = 20$ MN/m$^{3/2}$

A typical distribution of maximum shear strain in the vicinity of the crack tip is shown in Figure 5. Since metals fail principally by shear, the maximum shear stress is thought to be the most relevant parameter. There is often near symmetry of γ_{max} relative to the crack plane, in contrast to the displacement diagram and crack opening displacement. The normal and principal strains are usually nonsymmetric, and present patterns of strain which are difficult to interpret. The effective total strain

$$\varepsilon_t^{eff} = \frac{2}{\sqrt{3}} \left(\varepsilon_1^2 + \varepsilon_1 \varepsilon_2 + \varepsilon_2^2 \right)^{1/2}$$

has also been examined because it combines the principal strains; it is usually found to behave much like the maximum shear strain, although not always, and is easier to interpret than the principal strains alone.

A method for computing the stress tensor from the strain tensor has been found [12], which, in addition to deriving stresses, has the advantage of allowing the separation of total strain into elastic and plastic components. This computational method, of course, also required the use of a constitutive equation; the cyclic stress-strain curve is used. Maximum effective stress at the crack tip, as determined by these computations,

is in the range of 2.3 to 3.5 times the yield stress [12,13], which correlates reasonably well with finite element computations [14]. The stress
distribution is much the same as the strain distribution, such as is
illustrated in Figure 5, except that the distribution peaks much less at
the crack tip due to the relatively low work hardening exponent of the
materials studied.

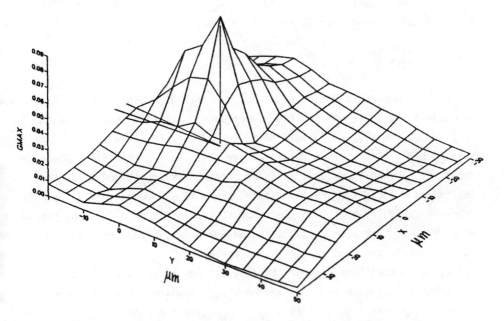

Figure 5 - Distribution of Maximum Shear Strain for Ti-6Al-4V at
$\Delta K = 9.4$ MN/m$^{3/2}$ in Vacuum. Crack is shown schematically
on zero strain plane. x and y are in micrometers.

Stress computations have proved to be the most useful in efforts to
understand overloads and the subsequent retardation of crack growth which
follows an overload. By analyzing the individual steps in the single overload loading and unloading sequence, the residual stresses caused by this
process may be computed. These computations have proved to be very
difficult, and the subject of fatigue crack growth subsequent to the
overload is still under investigation.

4. MODELLING FATIGUE CRACK BEHAVIOUR

Because fatigue crack growth has been found to be periodic, the old
concept of treating the material at the crack tip as a low cycle fatigue

specimen has been utilized, and the material failure is considered to be described by the equation

$$\Delta\epsilon_p \, \Delta N_c^{\beta} = \epsilon_c \, , \tag{1}$$

where $\Delta\epsilon_p$ = the strain experienced by the failing element, ΔN_c = the number of cycles for failure and ϵ_c = the cumulative strain for failure. Substituting ΔN of equation (1) into $\Delta a/\Delta N$ gives

$$\frac{da}{dN} = \frac{\Delta a}{\Delta N} = \Delta a \left(\frac{\Delta\epsilon_p}{\Delta\epsilon_c}\right)^{1/\beta} . \tag{2}$$

The crack tip plastic strain range $\Delta\epsilon_p$ has been determined for the ingot aluminum alloy 7075-T651 and the powder metallurgy alloy MA-87 (now 7091) in air containing water vapour and in very dry nitrogen or vacuum. These results have been reported fully elsewhere [9,10]. Considering only crack growth without environmental influence, the crack growth model may be derived as follows [9]. Strain at the crack tip is considered to be the "damage" caused by cyclic loading, and ΔK is used as a correlating parameter which assumes that similitude exists between the various crack length-stress level combinations used in the investigation.

Lines statistically fit to the data for crack tip plastic strain range are shown in Figure 6 for 7075 and MA-87. These lines are described by equation (3), where $\Delta K_{eff} = \Delta K - \Delta K_{TH}$; ΔK_{TH} = threshold ΔK,

$$\Delta\epsilon_p = K_0 \, \Delta K_{eff}^{r} . \tag{3}$$

Striation spacing has also been measured for these materials, and is considered to be the growth increment. Striation spacing is correlated with ΔK_{eff} as [15]

$$\Delta a = A_0 \, \Delta K_{eff}^{n} . \tag{4}$$

Crack growth rate is also a function of ΔK_{eff}, and is given by

$$\frac{da}{dN} = B \, \Delta K_{eff}^{s} . \tag{5}$$

Combining equations (3-5), with equation (2) allows ϵ_c to be derived as

$$\epsilon_c = \left(\frac{A_0}{B}\right)^{\beta} K_0 \, \Delta K_{eff}^{(n+(r/\beta)-s)\beta} . \tag{5}$$

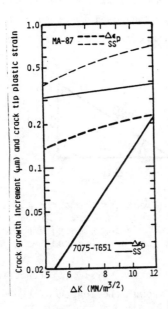

Figure 6 - Correlation of the Measured Striation Spacing (SS), which is the same as the Crack Growth Increment Δa, and the Crack Tip Plastic Strain $\Delta \varepsilon_p$

By assuming equation (1) represents a failure criterion independent of ΔK_{eff}, it is possible to solve for both ε_c and β. For 7075, values of β and ε_c so derived correlate very well with values as derived by low cycle fatigue tests. For MA-87, the agreement is not nearly so good, but the low-cycle fatigue data with which to compare are sparse.

A geometric model of the crack tip which related crack tip opening displacement (CTOD), the growth increment and the measured strain has also been devised [16]. The model is a variation of the BCS model concept in that it assumed that CTOD is caused by slip on discrete slip planes. Slip bands are limited in length by the microstructure of the material, so in this respect the model is similar to other blocked slip band models. The number of slip bands at the crack tip and the angle between the slip direction and direction of crack growth are governed by CTOD and Δa; this model differs from others in this respect. CTOD has also been measured for the aluminium alloys, and is described by the equation

$$CTOD = C_0 \Delta K^q . \tag{7}$$

The measured strain, equation (3) is used to compute the stress on the
slip plane pileup, which is then used to compute the slip offset $D_N b$, whi‹
is related to CTOD and Δa, Figure 7, and to r_s, the length of the slip li

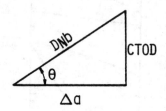

Figure 7 - Elements in the Crack Tip Geometric Model. $D_N b =$ *slip offset*
on blocked slip band, Δa = *crack growth increment (striation*
spacing) and CTOD = *crack opening displacement 1 μm behind*
tip. The angle θ *is set by these measured values.*

This model may be used in two ways: (1) if striation spacing, CTOD
and crack tip strains are known, the length of the slip band r_s, the
angle between the slip band and growth direction θ, and the number of sli›
bands may be computed, allowing a microstructural element in the material
to be related to r_s.   This computation has been done for 7075-T651, where
the derived r_s correlated well with the mean free path for dislocation
slip between dispersoids; (2) if CTOD and $\Delta\varepsilon_p$ are measured and knowledge
of the microstructure is used to assume a value of r_s, then, with a few
assumptions, it is possible to compute Δa, the crack growth increment.
Crack growth rate may then be determined using equation (1) together with
low cycle fatigue data.

The model has been tested in this second way for Ti-6Al-4V in the RA
condition [17], which has a microstructure that allows a good estimate of
r_s.   The values of Δa derived using the model were then verified by measuring striation spacings from the fracture surface, with good agreement
obtained.

With these models, it appears to be possible to utilize microstructura‹
information, together with measured crack tip opening and strain parameter‹
and low cycle fatigue data, to predict the crack growth rate which would
be expected.   The usefulness of these models, however, will be rather to
consider changes in microstructural parameters which would alter the fatig‹
crack growth resistance of the material.

The models briefly mentioned above have been developed as working hypotheses and in order to try to see how various measured parameters might be connected. This is but one step in the deeper understanding of fatigue crack growth. Specifically, these models do not consider a number of factors, such as crack closure and stress state, which are known to be related to fatigue crack growth. Thus, these models need to be further modified and evolved.

5. SUMMARY AND CONCLUSIONS

(1) Dynamic high-resolution observation of fatigue crack growth at low ΔK_{eff} has indicated that crack growth in intermittent.

(2) Crack opening displacement is generally a mix of Mode I and Mode II.

(3) Crack tip opening displacement increases with increasing ΔK.

(4) Crack tip strains have been determined using the stereoimaging technique, and are found to increase with increasing ΔK.

(5) These observations of crack tip behaviour have been included in two models. One which treats the crack tip element as a low cycle fatigue specimen, and the other which geometrically relates the various measured crack tip parameters with microstructural factors.

ACKNOWLEDGEMENTS

The author is grateful to the Air Force Office of Scientific Research, the Office of Naval Research and Southwest Research Institute for the continued support necessary to perfect and use the tools required for this research. My long professional collaboration with Dr. James Lankford during the course of this research is also especially acknowledged.

REFERENCES

1. DAVIDSON, D.L. and NAGY, A., *J. Phys. E.*, Vol. 11, 1978, pp. 207-210.

2. DAVIDSON, D.L., *Scanning Electron Microscopy*, Vol. II, 1979, pp. 79-86.

3. WILLIAMS, D.R., DAVIDSON, D.L. and LANKFORD, J., *Experimental Mechanics*, Vol. 20, 1980, pp. 134-139.

4. DAVIDSON, D.L. and LANKFORD, J., "High-Strength Powder Metallurgy Aluminum Alloys", *TMS-AIME*, 1982, pp. 163-176.

5. DAVIDSON, D.L., "Micro- and Macro-Mechanics of Crack Growth", *TMS-AIME*, 1982, pp. 161-176.

6. LANKFORD, J., *Fatigue of Eng. Mat. and Struct.*, Vol. 5, 1982, pp. 233-248.

7. DAVIDSON, D.L. and LANKFORD, J., "Fatigue Mechanisms", *ASTM STP 675*, 1978, pp. 277-284.

8. NIX, K.J. and FLOWER, H.M., *Acta Met.*, Vol. 30, 1982, pp. 1549-1559.

9. DAVIDSON, D.L. and LANKFORD, J., *Fat. Eng. Mat. and Struct.*, Vol. 4, 1983, pp. 287-303.

10. DAVIDSON, D.L. and LANKFORD, J., *Fat. Eng. Mat. and Struct.*, in press.

11. LANKFORD, J. and DAVIDSON, D.L., *Met. Trans.*, Vol. 14A, 1983, pp. 1227-1230.

12. DAVIDSON, D.L., WILLIAMS, D.R. and BUCKINGHAM, J., *Experimental Mechanics*, Vol. 23, 1983, pp. 242-248.

13. DAVIDSON, D.L. and LANKFORD, J., *Int. J. Fracture*, Vol. 17, 1981, pp. 257-275.

14. LEVY, N., MARCAL, P.V., OSTERGREN, W.J. and RICE, J.R., *Int. J. Fracture Mechanics*, Vol. 7, 1971, pp. 143-156.

15. LANKFORD, J. and DAVIDSON, D.L., *Acta Met.*, Vol. 31, 1983, pp. 1273-128

16. DAVIDSON, D.L., *Acta Met.*, in press.

17. DAVIDSON, D.L. and LANKFORD, J., *Met. Trans.*, in preparation.

MODELLING PROBLEMS IN CRACK TIP MECHANICS   *Waterloo, Ontario, Canada*
CFC10, University of Waterloo   *August 24-26, 1983*

CRACK TIP GROWTH RATE MODEL FOR CYCLIC LOADING

F. ELLYIN and C.O.A. FAKINLEDE
Department of Mechanical Engineering
University of Alberta
Edmonton, Alberta, Canada

1. CRACK GROWTH MODEL

The material ahead of the crack tip is modelled as an assemblage of
uniaxial fatigue elements [1]. It is the behaviour of these material
elements which predicts the crack propagation rate for any given configuration. Fatigue damage leads to failure when the total amount of damage
reaches a critical value. Each fatigue element is subjected to increasing
levels of stress-strain fields as the crack tip moves towards it. The
linear damage accumulation rule proposed by Miner [2] is used to evaluate
the failure under this variable loading condition. The failure of an
element indicates a crack initiation at the material position it occupies,
and this sequence of initiations is perceived as stable crack growth on
the macroscopic scale. The damage region considered is the cyclic plastic
zone ahead of the crack tip.

As the general equations for the stress/strain fields ahead of a
fatigue crack are not known, approximations are made from the stationary
solutions as suggested by Rice [3]. For the strain hardening material, it
often happens that fatigue cycling results in stable hysteresis loops
after the application of few cycles [4]. In such an event, one may reasonably assume that the loading and unloading curves are identical, thus
the unloading curve may be referred to a set of reversed axes from the
point of unloading. The plastic superposition procedure suggested by
Rice [3] can then be used to obtain the fatigue stress/strain fields.
While this procedure is tenable for a large portion of the damage zone,
it has been widely recognized (Majumdar and Morrow [5], Glinka [6], Shih
et al [7]) that a small region exists immediately adjacent to the crack
tip where, due to nonproportional plasticity, the stress/strain field
cannot be found in this way. This region (depicted as region I) is treated
differently from the rest of the damage zone (region II). In the latter,

proportional plastic deformation will be assumed. In the former, since
the deformation is highly nonproportional, the fatigue fields have to be
found in another way. However, bounds on the stress/strain field can be
estimated from experimental considerations. In region II (Figure 1) the
stationary singularity solution of Hutchinson [8] is adopted as described
in [3]. The normal stress/strain range at any point x ahead of a tensile
crack in a material modelled by the generalized Ramberg-Osgood stress-
strain law [6,8] are:

$$\Delta\varepsilon = \frac{\sigma_y'}{E}\left((B-\nu A)\left(\frac{\Delta K^2}{\alpha I\sigma_y'^2 x}\right)^{1/(n+1)} + \left(B - \frac{A}{2}\right)\alpha^{1/(n+1)}\left(\frac{\Delta K^2}{I\sigma_y'^2 x}\right)^{n/(n+1)}\right),$$

$$\Delta\sigma = \sigma_y'\left(B\left(\frac{\Delta K^2}{\alpha I\sigma_y'^2 x}\right)^{1/(n+1)}\right),$$

(1)

where α, ν and n are material constants, σ_y' is the material yield stress,
and K is the Stress Intensity Factor. The constants A and B depend on
the loading conditions while I is the integral function of the hardening
parameter n. Figure 1 shows a typical fatigue element as the crack moves
towards it. Taking the crack-tip as origin, this is equivalent to the
element moving through the damage zone. Let the damage zone be divided
into small subregions as shown, the number of cycles spent in the interval
x_i, x_{i+1} is given by

$$\frac{x_{i+1}-x_i}{(da/dN)_i} \equiv \frac{\Delta x_i}{(da/dN)_i}.$$

(2)

Here $(da/dN)_i$ is the mean crack growth rate in this interval. If N_{fi} is
the number of cycles to failure under combined effect of the stress and
strain ranges, the cycle ratio

$$\frac{\Delta x_i/\left(\frac{da}{dN}\right)_i}{N_{fi}}$$

(3)

represents the ratio of the damage accumulated in the range x_i, x_{i+1} to
the total damage to failure. Hence

$$\sum_{\text{region II}} \frac{\Delta x_i/\left(\frac{da}{dN}\right)_i}{N_{fi}} = p,$$

(4)

where $0 < p < 1$ is the proportion of damage accumulated in region II. Equation (4) expresses the fact that the fatigue element does not fail in region II, but that it is weakened by the amount of damage undergone there. If region II is small compared to the crack-length, then $(da/dN)_i$ is approximately independent of i, so that equation (4) becomes:

$$p \frac{da}{dN} = \sum_{II} \frac{x_i}{N_{fi}} . \qquad (5)$$

Thus if N_{fi} can be found for all x_i in this region, the knowledge of the "damage ratio" p is all that is needed to find the crack growth rate.

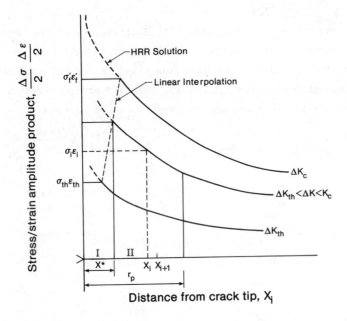

Figure 1 - Stress/Strain Range Distribution Ahead of the Crack Tip

The damage ratio is however difficult to determine. Its value depends on the size ratio of the two regions as well as the cyclic stress/strain distributions. It therefore may not be a constant. An indirect approach will be used to find the growth rate. This is achieved by examining the damage due to region I.

The premise here is that for stable crack growth, the maximum cyclic stress/strain sustainable by the fatigue elements cannot exceed the fatigue-strength and ductility limits of material σ_f', ε_f', respectively. Values

of stress/strain higher than these, when they exist, would lead to an
increasing sequence of crack growth. Such finite growths occurring a
number of times in less than one cycle is evidently unstable growth. Suf
ficient time is not allowed for a stable translation of the stress/strain
front with the crack. An upper bound for the cyclic stress/strain field
in region I is thus given by the strength and ductility coefficients. Th
bound is evidently attainable only at the transition values of the applie
stress intensity factor range. In this case, from equation (1), neglecti
elastic terms, the maximum extent of region I is given by:

$$x^* = C \, \frac{\Delta K_c^2}{4 \pi \sigma_y'^2 \varepsilon_f' \sigma_f'} \tag{6}$$

where ΔK_c is the critical stress intensity factor range of the material,
while C is a constant which can be evaluated from the material constants
in (1). The stress/strain field in region I can be approximated by the
values obtained from equation (1) for the point x^* provided that x^* is
small. Using (3) for this region,

$$\frac{x^* / \frac{da}{dN}}{N_f(x^*)} = 1 - p , \tag{7}$$

and, from equations (5) and (7),

$$\frac{da}{dN} = \frac{x^*}{N_f(x^*)} + \sum_{II} \frac{\Delta x_i}{N_{fi}} , \tag{8}$$

in the limit that i tends to infinity, i.e.,

$$\frac{da}{dN} = \frac{x^*}{N_f(x^*)} + \int_{x^*}^{r_p} \frac{1}{N_f(x)} \, dx \tag{9}$$

is the proposed crack propagation law. It is noted here that the two
extreme values of p (i.e., 0 and 1) specialize equation (8) to that of
Glinka [6] and Majumdar and Morrow [5]. The numerical computations here
however show that neither is a realistic estimate of the damage ratio.
The upper limit of integration r_p is the extent of the reversed plastic
zone.

The expressions relating stress/strain ranges to the number of cycles to failure in a strain controlled experiment [4] are:

$$\frac{\Delta\varepsilon}{2} = \frac{\sigma_f'}{E} N_f^b + \varepsilon_f' N_f^c , \qquad \frac{\Delta\sigma}{2} = \sigma_f' N_f^b , \qquad (10)$$

where b and c are material constants. Equations (10) can be multiplied to give:

$$\Delta\sigma\Delta\varepsilon = \frac{4\sigma_f'^2}{E} N_f^{2b} + 4\varepsilon_f'\sigma_f'N_f^{b+c} , \qquad (11)$$

which is used to find the number of cycles to failure of a fatigue element at any given stress/strain range.

The size of region I can be estimated for the two extreme values of the external loading. When both threshold and critical values of the stress intensity factor range are known, the values of x* for an intermediate value of the loading can be obtained, for example, by a linear interpolation. This linear interpolation is valid as long as the two extreme values are of the same order or magnitude. In other cases, this is a first approximation only.

2. COMPARISON WITH PUBLISHED EXPERIMENTAL DATA

Since a closed form inversion of the low-cycle fatigue equation (11) cannot be found, an iterative Newton-Raphson algorithm was used. An eight-point Gaussian quadrature formula was also used to evaluate the integral in (9). The value of x* obtained from equation (6) was used for all levels of the stress intensity factor range. It is more accurate to do an interpolation between the threshold and the critical values of the stress intensity factor range when both are available. The two terms on the LHS of equation (9) were found to be of similar magnitude for most values of external loadings (i.e., ΔK_c) shown in Figures 2 - 5.

Figures 2 - 5 are the plots of the theoretical predictions and the experimental data for three types of steels and aluminium. Data in Figures 4 and 5 are taken from Saxena and Hudak [9] while that in Figure 2 are from Barsom [10]. Data in Figure 3 are from Ellyin and Li [11]. Material properties used for the figures shown are given in Table 1. It is noted that there is fair agreement with experimental data.

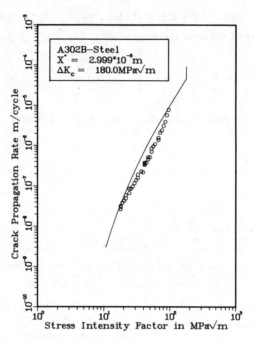

Figure 2 - Theoretical Predictions and Experimental Data of Fatigue Crack Growth Rate for A302B-Steel

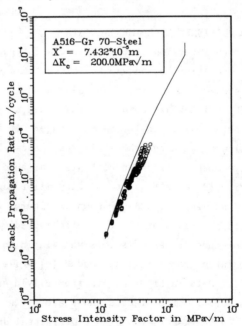

Figure 3 - Theoretical Predictions and Experimental Data of Fatigue Crack Growth Rate for A516 Gr 70-Steel

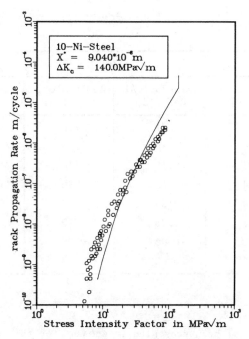

Figure 4 - Theoretical Predictions and Experimental Data of Fatigue Crack Growth Rate for 10-Ni-Steel

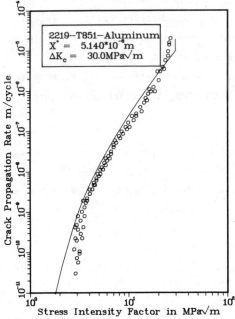

Figure 5 - Theoretical Predictions and Experimental Data of Fatigue Crack Growth Rate for 2219-T851-Aluminum

Table 1 - Material Properties used to Calculate the Crack Growth Rate in Equation (1)

Material Property	Alloy			
	10-Ni-Steel Ref. [9]	2219-7851 Al Ref. [9]	A302B Steel Ref. [10]	A516-Gr70 Steel Ref. [11]
E MPa	206	71	206	200
σ'_y MPa	1200	330	300	310
ε'_f	0.56	0.35	0.48	0.26
σ'_f MPa	1634	613	950	900
ΔK_c MPa$\sqrt{m}$	140	30	180	200
n'	9.17	8.26	5.26	5.26
b	-0.08	-0.08	-0.08	-0.11
c	-0.65	-0.55	-0.60	-0.51

3. CONCLUSIONS

A new approach has been presented to account for the crack growth by fatigue damage accumulation in the entire damage zone. The theoretical singularity at the crack tip is avoided in the cyclic loading in a way consistent with experimental observation. The damage contribution in both proportional and nonproportional plastic deformation zones are found to be of similar order of magnitude, for a wide range of stable crack growth. Neglecting the contribution of either part (as in References [5,6]) may introduce error in predicting the crack growth rate.

4. ACKNOWLEDGEMENT

The work reported here is part of a general investigation into material properties under multiaxial states of stress. The research is supported, in part, by the Natural Sciences and Engineering Research Council of Canada (NSERC Grant No. A-3808).

REFERENCES

1. HEAD, A.K., "The Growth of Fatigue Cracks", *Philosophical Magazine*, Vol. 44, 1953, pp. 925-938.
2. MINER, M.A., "Cumulative Damage in Fatigue", *J. of Applied Mechanics, ASME*, Vol. 12A, 1945, pp. 159-164.
3. RICE, J.R., "Mechanics of Crack Tip Deformation and Extension by Fatigue", *ASTM STP 415*, 1967, pp. 247-311.

4. LEFEBVRE, D. and ELLYIN, F., "Cyclic Response and Inelastic Strain Energy in Low-Cycle Fatigue", *Int. J. of Fatigue,* Vol. 6, No. 1, 1984.

5. MAJUMDAR, S. and MORROW, J., "Correlation between Fatigue and Low Cycle Fatigue Properties", *Fracture Toughness and Slow Stable Cracking,* *ASTM STP 559,* 1974, pp. 159-182.

6. GLINKA, G., "A Cumulative Model of Fatigue Crack Growth", *Int. J. of Fatigue,* Vol. 4, 1982, pp. 59-67.

7. SHIH, C.F., et al, "Studies on Crack Initiation Stable Crack Growth", *ASTM STP 668,* 1979, pp. 65-120.

8. HUTCHINSON, J.W., "Singular Behaviour at the End of a Tensile Crack in a Hardening Material", *J. of the Mechanics and Physics of Solids,* Vol. 16, 1968, pp. 13-31.

9. SAXENA, A. and HUDAK, S.J., "Role of Crack-Tip Stress Relaxation in Fatigue Crack Growth", *Fracture Mechanics, ASTM STP 677,* 1979, pp. 215-232.

10. BARSOM, J.M., "Fatigue Crack Propagation in Steels of Various Yield Strengths", *J. of Engineering for Industry, ASME,* Vol. 93, 1971, pp. 1190-1196.

11. ELLYIN, F. and LI, H-P., "Fatigue Crack Growth in Large Specimens with Various Stress Ratios", *J. of Pressure Vessel Technology,* *Trans. ASME,* to appear.

MODELLING PROBLEMS IN CRACK TIP MECHANICS   *Waterloo, Ontario, Canada*
CFC10, University of Waterloo   *August 24-26, 1983*

A STOCHASTIC APPROACH TO THE GROWTH AND COALESCENCE OF
MICROVOIDS IN A DUCTILE SOLID

Y. HADDAD
Department of Mechanical Engineering
University of Ottawa
Ottawa, Ontario, Canada

1. INTRODUCTION

The fracture of ductile solids has been frequently referred to [1] as
fracture phenomena that involve the following three essential events:
(i) nucleation of voids at inclusions, (ii) growth of voids under further
straining and (iii) interconnection of voids leading to final fracture.

In a conventional approach to the study of the problem, continuum and
modified continuum models have been developed by a large number of authors.
Although considerable progress has been made in the understanding of various
events which contribute to ductile failure, no ductile failure theory, as
yet, exists. Due to the existence of defects in the material and when
these defects become significant, the field quantities fail to be continuous
at the separating boundaries. Thus, it becomes necessary to consider these
local effects as an integral part of the problem. In this context, the
need for a probabilistic approach to the problem, that accounts for the
random nature of the microstructure, has been discussed quite frequently
in the literature, [2].

2. APPROACH

In the present analysis, the material system is regarded as heterogeneous
medium of actual microstructural elements. These elements may exhibit random
geometric and physical characteristics and are further disturbed by the
nucleation of a latitude of randomly oriented microvoids or fissures [3,4].
In dealing with such a system, the significant field quantities are considered as random variables and the deformation process is seen as a
stochastic process.

An element of the medium is defined as the smallest region of the microstructure that represents the mechanical and physical properties of the
material at a local level. Before crack initiation, the individual element

may be regarded as continuous, hence, the state of the local strain may be
determined by a continuum constitutive equation. At this state, the local
elastic-plastic response may be formulated on the basis of a generalized
yield function that includes both the anisotropic work hardening and
Bauschinger effect, [5]. In a defective element, however, the growth of
a microvoid or fissure may be assumed to follow a random walk of the discrete Markov type [6]. The latter may be associated with the maximum strain
ahead of the crack tip and could be expressed in terms of the build-up
of strain due to preceding crack growth increment, growth of interacting
cracks and the redistribution of local stress. Two probabilities of absorp-
tion could be involved here, i.e., at the structural element boundary and
at ∞. The so-called "simultaneous coalescence" and "eventual coalescence"
between two neighbouring fissures are introduced and correlated to the
characteristics of the local microstructure.

As the crack reaches the boundary between two neighbouring structural
elements, an inter-elemental fracture may set-up. In this case, a stationary Markov process [7] model will be examined in relation to the grain
boundary structure and the associated binding potential between the two
elements [8].

Since the material system that occupies a given physical domain is
regarded in the proposed research as a discrete medium, a transition from
the local or discrete description to the macroscopic one must be attempted.
In this context, the concept of the intermediate domain [7] in the material
is utilized. Here, the discrete fracture process through a statistical
ensemble of structural elements will be considered on the basis of a
stochastic model with transformation probabilities. The latters are dependent on the kinetics of the surrounding microstructure.

3.   LOCAL RESPONSE BEHAVIOUR

A structural element (α) of the material is chosen to represent the
response behaviour of an individual grain, as well as the bonding effect
within the grain boundary of two neighbouring grains.

3.1 Deformation Kinematics

The deformation kinematics of an element (α) is shown in Figure 1.
Two reference frames are used, i.e., a body frame $^{\alpha}Y_1$, $^{\alpha}Y_2$, $^{\alpha}Y_3$ attached
to the centre of the grain with $^{\alpha}Y_1$ coincident with its major dimension,

and an external frame Z_1, Z_2, Z_3. An additional frame at the nucleation centre of the crack, that can be used to describe the orientation of the major dimension of the crack within the grain, is $^{\alpha}X_1$, $^{\alpha}X_2$, $^{\alpha}X_3$.

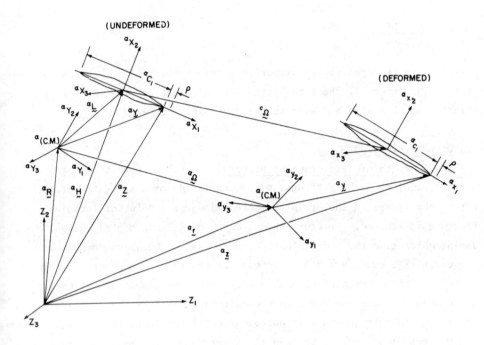

Figure 1 - Kinematics of Crack Growth in an Element "α"

3.2 Continuous Grain (State of Strain Before Crack Nucleation)

Based on a generalized yield function f ($\xi^{\lambda\mu}$, $\varepsilon_{\lambda\mu}$) for the anisotropic state resulting from the strain history $\varepsilon_{\lambda\mu}$ and including the Bauschinger effect, a constitutive equation for the continuous grain may be written, in an operational form as:

$$\xi^{vk} = \Gamma_{vk\lambda\mu} D \varepsilon_{\lambda\mu} \, , \tag{1}$$

where the material operator $\Gamma_{vk\lambda\mu}$ takes the form:

$$\Gamma_{vk\lambda\mu} = \left[E_{vk\lambda\mu}^{-1} + \left\{ g'_{\lambda\mu vk} + A \, (L'_{\lambda\mu vk} - \frac{2}{3} g_{\lambda\mu} \, \varepsilon_{vk}) \right. \right.$$
$$\left. \left. + A^2 \, (M'_{\lambda\mu vk} - \frac{1}{3} M'_{\ell m vk} \, g^{\ell m} \, . \, g_{\lambda\mu}) \xi'_{vk} - B \varepsilon_{\lambda\mu} \right\} \frac{D_\lambda}{\xi'vk} \right]^{-1} \, , \tag{2}$$

in which:

$$L'_{\lambda\mu\nu k} = \frac{1}{2} (g_{\lambda\nu}\varepsilon_{\mu k} + g_{\mu k}\varepsilon_{\lambda\nu} + g_{\mu\nu}\varepsilon_{\lambda k}) \ ,$$

$$M'_{\lambda\mu\nu k} = \frac{1}{2} (\varepsilon_{\lambda\nu}\varepsilon_{\mu k} + \varepsilon_{\lambda k}\varepsilon_{\mu\nu}) \ ,$$

$$D\lambda = \frac{Df}{F' \ \xi^{\lambda\mu} \ \dfrac{\partial f}{\partial \xi^{\lambda\mu}}} \ ,$$

where $g_{\lambda\nu}$ is the spherically symmetric metric tensor, $F' = DF(W)/DW$ (W is the plastic work), the terms with A and A^2 represent the anisotropy effects and that with B designate the Bauschinger effect.

3.3 Defective Grain

3.3.1 Growth of Transgranular Fissure. In this paper, the growth of a transgranular crack is thought of as consisting of a series of steps. First, the stress on the grain is increased until a point of incipient fracture is reached. Fracture then occurs for an incremental distance during which time the redistribution of stress in the neighbourhood of the crack will cause a further increase in the total accumulated strain there. If this increased strain due to redistribution of the stress is not enough to leave the grain in a condition to satisfy the fracture criterion, further growth will not occur until the local stress is increased. The process of increasing the applied stress and then cracking with associated straining and redistribution of stress is repeated stochastically until a stage is reached in which there is enough strain to satisfy the fracture instability criterion. Under these conditions, no further increase in the local stress might be required and the crack will become unstable to reach the grain boundary.

Thus, with reference to Figure 2, the growth of a transgranular fissure is assumed to follow a random walk of a finite set of states X = 0,1,..., i,..., a where X = a is an absorbing barrier representing the grain boundary ahead of the crack tip. Two probabilities p and q may be introduced:

  p: the probability that after the first step, the crack will grow
     from the position i to i+1.

  q: the probability that after the first step, no growth of the
     crack will occur.

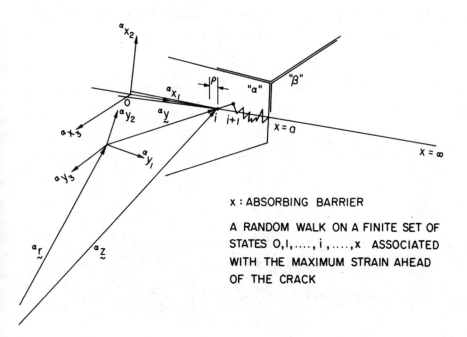

x : ABSORBING BARRIER

A RANDOM WALK ON A FINITE SET OF
STATES 0,1,....,i,....,x ASSOCIATED
WITH THE MAXIMUM STRAIN AHEAD
OF THE CRACK

Figure 2 - Growth of a Single Transgranular Fissure

In view of the discussion above, the probability p may be defined as
(see Figure 1):

$$p = P\left\{\left(\frac{\partial \varepsilon_{\lambda\mu}}{\partial y_\lambda}\right)_{c_\lambda, \xi_{\lambda\mu}} dc_\lambda - \left(\frac{\partial \varepsilon_{\lambda\mu}}{\partial c_\lambda}\right)_{\xi_{\lambda\mu}} dc_\lambda \le \left(\frac{\partial \varepsilon_{\lambda\mu}}{\partial \xi_{\lambda\mu}}\right)_{c_\lambda, y_\lambda} d\xi_{\lambda\mu}\right\}, \tag{3}$$

where the first term on the RHS of (3) represents the plastic strain
gradient at a unit distance in front of the current tip of the crack,
the second term represents the plastic strain gradient occurring during
the previous element growth and the last term designates the increase in
the stress that would be necessary to provide for the difference between
the first two terms.

Let $\Pi_{i,\mu}$ denote the probability that the random walk of the crack tip
will terminate with the μth increment of time at the barrier (a) between
the two joining elements α and β when the initial position of the crack
tip is i.

At the first step, the position will be either $i + 1$ or i with probabilities p and $q = 1-p$ respectively. Hence, for $1 < i < a-1$ and $\mu \geq 1$, one may write the differential equation:

$$\Pi_{i,\mu+1} = p\,\Pi_{i+1,\mu} + q\Pi_{i,\mu} , \tag{4}$$

subject to the boundary conditions

$$\Pi_{0,\mu} = 0 , \quad \Pi_{a,\mu} = 0 \quad \text{for} \quad \mu > 1 ,$$

$$\Pi_{0,0} = 1 , \tag{5}$$

$$\Pi_{i,0} = 0 \quad \text{for} \quad i > 1 .$$

The solution of (4) subject to the boundary conditions (5) has been dealt with in [6], and may be written as:

$$\Pi_{i,\mu} = a_{2\mu}^{-1}\, p^{(\mu-1)/2} q^{(\mu+1)/2} \sum_{k=1}^{a-1} \cos^{\mu-1}\left(\frac{\Pi k}{a}\right) \sin\left(\frac{\Pi k}{a}\right) \sin\left(\frac{\Pi i k}{a}\right) . \tag{6}$$

On the other hand, there is the possibility that the crack extension would not stop at the grain boundary and continue to grow with a barrier at $X = \infty$. In this case, equation (6) becomes:

$$\Pi_{i,\mu} = 2^{\mu} p^{(\mu-i)/2} q^{(\mu+i)/2} \int_0^1 \{\cos^{\mu-1}(\Pi x)\, \sin\,(\Pi x)\, \sin\,(\Pi x i)\}\, dx , \tag{7}$$

which satisfies that the crack will either continue to grow without limit or not depending on p and q.

3.3.2 <u>Two Interacting Fissures</u>. In Figure 3, the local orientations of two neighbouring cracks are identified by the two local coordinate frames X_1, X_2, X_3 and X_1', X_2', X_3' where, as shown X_1 designates the major dimension of the first crack and X_1' identifies the major dimension for the second one. The axes X_1 and X_1' may intersect, as shown in Figure 3, at a point a_1. The latter will be also a common point in the two crack affected zones formed in front of the two propagating cracks.

Thus, the tip of the first crack may follow, during the fracture process, a random walk characterized by a finite set of states $X = 0, 1, \ldots, i$, $\ldots, a_1$ with probabilities p, q and $\Pi_{i,\mu}$ (Section 3.3.1). The second fissure may, also, follow in its growth a random walk defined by the finite set of states $X' = 0, 1, \ldots, i', \ldots, a_1$ with probabilities p', q' and $\Pi_{i',\mu'}$ where $p' = p'(p)$ and $q' = q'(q)$. Accordingly, a simultaneous coalescence of the two fissures may occur at the barrier $X = a_1$ if:

$$\Pi_{i,\mu} = \Pi_{i',\mu'} , \tag{8}$$

subject to the conditions (5).

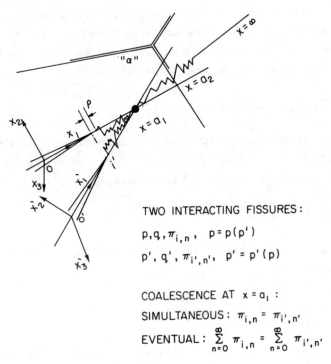

TWO INTERACTING FISSURES:

$p, q, \pi_{i,n}$, $p = p(p')$

$p', q', \pi_{i',n'}$, $p' = p'(p)$

COALESCENCE AT $x = a_1$:

SIMULTANEOUS: $\pi_{i,n} = \pi_{i',n'}$

EVENTUAL : $\sum\limits_{n=0}^{\infty} \pi_{i,n} = \sum\limits_{n=0}^{\infty} \pi_{i',n'}$

Figure 3 - Coalescence of Two Interacting Fissures

Further, if the two fissures do not coalesce at a_1, they may coalesce eventually at another barrier with a probability given by the condition:

$$\sum_{\mu=0}^{\infty} \Pi_{i,\mu} = \sum_{\mu'=0}^{\infty} \Pi_{i',\mu'} . \tag{9}$$

Once the two fissures coalesce, a common new tip will be formed that could start again a new random walk of crack growth.

As soon as the crack tip strikes, in its progress, the grain boundary ahead, a debonding affect might take place between the two adjoining grains.

It is, thus, important to include, on a local level, the grain boundary response behaviour. The debonding effect within the grain boundary will be treated, however, on a global scale when dealing (Section 4) with a statistical ensemble of structural elements within an intermediate domain of the material.

3.4 Grain Boundary

As far as the bonding effect, between two matching grains α and β, is concerned, the most suitable form can be expressed in terms of a Morse function [9]. The analytical form of this function which will represent the 3-dimensional case can be written as follows:

$$^{\alpha\beta}\psi = {}^{\alpha\beta}\psi_0 \{\exp(-2b|{}^{\alpha\beta}\underset{\sim}{d}|) - 2\exp(-b|{}^{\alpha\beta}\underset{\sim}{d}|)\} , \tag{10}$$

in which $^{\alpha\beta}\psi_0$ is the equilibrium interaction potential, b is the Morse constant and $^{\alpha\beta}\underset{\sim}{d}$ is the deformation in the bond. An expression for the interaction incremental stress can be formulated as follows:

$$\Delta\,^{\alpha\beta}\underset{\approx}{\xi} = {}^{\alpha\beta}\underset{\approx}{B}\,{}^{\alpha\beta}\underset{\sim}{d} , \tag{11}$$

where $^{\alpha\beta}\underset{\approx}{B}$ is a material tensor operator for the bonding interaction which takes the form [3,7]

$$^{\alpha\beta}\underset{\approx}{B} = \left[\frac{-2b^2\,{}^{\alpha\beta}\psi_0}{{}^{\alpha\beta}a}\,{}^{\alpha\beta}\underset{\sim}{n}\,{}^{\alpha\beta}\underset{\sim}{k}\,{}^{\alpha\beta}\underset{\sim}{k}{}^{-1}\right] , \tag{12}$$

where $^{\alpha\beta}a$ = area per bond, $^{\alpha\beta}\underset{\sim}{n}$ = outward unit normal to $^{\alpha\beta}a$, and $^{\alpha\beta}\underset{\sim}{k}$ is a unit base vector associated with the local coordinate frame.

4. TRANSITION TO MACROSCOPIC PHENOMENA

Here, the concept of the intermediate domain [9] in the material is introduced. It is established by the requirement that it is the smallest region of the medium on the boundary of which the macroscopic observables are still valid but, on the other hand, is large enough to contain a statistical number of structural elements. On the level of the intermediate domain, two fracture processes are considered:

(i) Growth of interacting fissures $j = 1, 2, \ldots, \ell$. Here the formulation of the probability of growth of one fissure (3) is generalized to include a group of interacting fissures.

(ii) Debonding of the grain boundaries. This process is seen to be stochastic of the stationary Markov type [7]. Hence, the process $\{\underset{\sim}{d}(t);$ $t > \mu\}$ is seen to be completely determined by its transition probabilities $\underset{\sim}{p}^d(s)$ where $s = t-\mu$.

REFERENCES

1. RICE, J.R. and TRACEY, D.M., "On the Ductile Enlargement of Voids in Triaxial Stress Fields", *J. Mech. Phys. Solids,* Vol. 17, 1969, pp. 201-217.
2. NASSER, N., (Editor), *Proc. of the IUTAM Symposium on Three-Dimensional Constitutive Relationships and Ductile Fracture,* Dourdan, France, 1980.
3. HADDAD, Y.M. and SOWERBY, R., "A Micro-Probabilistic Approach to the Ductile Deformation and Fracture of Metals - I. Preliminary Investigation", *Fracture 1977,* Vol. 2, ICF4, University of Waterloo, Canada, 1977, pp. 457-465.
4. HADDAD, Y.M., "A Three-Dimensional Constitutive Equation for a Ductile Solid with Randomly Oriented Fissures", *IUTAM Symposium,* Dourdan, France, 1980.
5. YOSHIMURA, Y. and TAKENAKA, Y., "Strain History Effects in Plastic Deformation of Metals", *Proc. IUTAM Symposium on Second Order Effects in Elasticity, Plasticity and Fluid Dynamics,* M. Reiner and D. Abir, Editors, The MacMillan Company, New York, 1964, pp. 729-750.
6. BHARUCHA-REID, A.T., *Elements of the Theory of Markov Process and Their Applications,* McGraw Hill, New York, 1960.
7. HADDAD, Y.M., *Ph. D. Thesis,* McGill University, Montreal, Canada, 1975.
8. PROVAN, J.W. and BAMIRO, O.A., "Elastic Grain-Boundary Responses in Copper and Aluminium", *Acta Metallurgica,* Vol. 25, 1977, pp. 309-319.
9. MORSE, P.M., "Diatomic Molecules According to the Wave Mechanics. II. Vibrational Levels", *Phys. Rev.,* Vol. 34, 1929.

MODELLING PROBLEMS IN CRACK TIP MECHANICS   *Waterloo, Ontario, Canada*
CFC10, University of Waterloo   *August 24-26, 1983*

ON THE INFLUENCE OF HIGH STRESS GRADIENTS ON LIGHT PROPAGATION
IN THE NEIGHBOURHOOD OF CRACK TIPS

F.W. HECKER
Technische Universität Braunschweig
Institut für Technische Mechanik, Experimentelle Mechanik
Braunschweig, Federal Republik of Germany

1. INTRODUCTION

Photoelastic methods in experimental strain or stress analysis are
based on the simplifying assumption, that both refracted rays are always
rectilinear and collinear. However it has been known for a long time that
in regions of high stress or strain gradients, both refracted rays are
curved and spatially separated. The resulting light deflections can
easily be measured [1]. This effect can be used to determine experimentally
coefficients of elastomechanic and photoelastic material equations, gradients
of particular components of stress tensor, etc. Methods and techniques
utilizing this effect have recently been named gradient photoelasticity
by Pindera [1]. On the other hand it is known that this effect limits the
resolution of isochromatic fringes in a polariscope [2,3,4] and that it
may influence the fringe order [5]. In this paper the approximate analysis
of light propagation inhomogeneous anisotropic bodies, outlined in [1],
is applied to the simple stress model of linear elastic fracture mechanics.
Estimated fringe order errors are discussed.

2. THEORETICAL ANALYSIS

The analysis is based on the following mathematical models [1]:

(1) linear elastic continuous body,

(2) geometrical optics,

(3) linearized photoelastic relations of Ramachandran and Ramaseshan.

As a sufficiently general basis the relations

$$n \frac{d^2x}{dz^2} = \left[\left(\frac{dx}{dz} \right)^2 + \left(\frac{dy}{dz} \right)^2 + 1 \right] \left(\frac{\partial n}{\partial x} - \frac{dx}{dz} \cdot \frac{\partial n}{\partial z} \right) , \tag{1}$$

$$n \frac{d^2y}{dz^2} = \left[\left(\frac{dx}{dz} \right)^2 + \left(\frac{dy}{dz} \right)^2 + 1 \right] \left(\frac{\partial n}{\partial y} - \frac{dy}{dz} \cdot \frac{\partial n}{\partial z} \right) , \tag{2}$$

have been chosen, where $\vec{x}$ = (x(z),y(z),z) is the position vector of the
considered ray in an arbitrarily chosen Cartesian coordinate system,
Figure 1.   The refractive index n = n(x,y,z,x',y') depends on the
ray direction (in the following d/dz is replaced by a prime).   Equations
(1) and (2) are used under the assumption, that the degree of load produced
anisotropy is so weak, that the wave normal and the tangential direction of
a ray practically coincide.

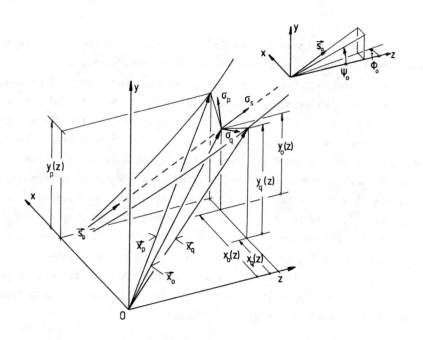

Figure 1 - Notations used for the Approximate Analysis of the Propagation
of Doubly Refracted Rays in Inhomogeneous Anisotropic Bodies

When the deviations of ray paths from rectilinearity are small, i.e.,
both rays deviate only slightly from the rectilinear path $\vec{x}_0$ = (x$_0$(z),y$_0$(z),
an incident ray would take in the unloaded body, Figure 1, the derivatives
of x,y occurring on the right sides of the equations (1), (2) may be approximated by

$$x' = x_0' = \tan \phi_0 , \qquad y' = y_0' = \tan \psi_0 . \qquad (3)$$

Simultaneously the refractive indices are replaced by the two secondary refractive indices n_p, n_q in the plane normal to the direction $\vec{s}_0$:

$$n_p = n_p(x,y,z,x_0',y_0') = n_0 + C_1\,\sigma_p + C_2(\sigma_q + \sigma_s)\,, \tag{4}$$

$$n_q = n_q(x,y,z,x_0',y_0') = n_0 + C_1\,\sigma_q + C_2(\sigma_s + \sigma_p)\,, \tag{5}$$

where σ_p and σ_q are the secondary principal stresses, σ_s is the normal stress in the direction of $\vec{s}_0$, and C_1, C_2 are the absolute photoelastic coefficients, depending on the wavelength. We arrive at

$$n_0\,x_p'' = (\tan^2\phi_0 + \tan^2\psi_0 + 1)\left(\frac{\partial n_p}{\partial x} - \tan\phi_0\,\frac{\partial n_p}{\partial z}\right), \tag{6}$$

$$n_0\,y_p'' = (\tan^2\phi_0 + \tan^2\psi_0 + 1)\left(\frac{\partial n_p}{\partial y} - \tan\psi_0\,\frac{\partial n_p}{\partial z}\right), \tag{7}$$

for one of the two related rays. The corresponding relations for the second ray are given by using subscript q instead of p. n_0 is the refractive index of the unloaded body; the partial derivatives $\partial n_p/\partial x$ etc., have to be taken at the actual locus $(x_0(z),y_0(z),z)$. As an example the equations (6), (7) have been applied in [5] to the case of a toughened glass plate at oblique incidence.

In case of normal incidence on a plane stress state, ϕ_0, ψ_0 and $\partial n/\partial z$ are zero, the secondary principal stresses σ_p, σ_q are replaced by the principal stresses σ_1 and σ_2, and $\sigma_s = \sigma_z = 0$. This leads to the relations

$$n_0 x_1'' = \frac{\partial n_1}{\partial x} = C_1\frac{\partial\sigma_1}{\partial x} + C_2\frac{\partial\sigma_2}{\partial x}\,, \tag{8}$$

$$n_0 y_1'' = \frac{\partial n_1}{\partial y} = C_1\frac{\partial\sigma_1}{\partial y} + C_2\frac{\partial\sigma_2}{\partial y}\,, \tag{9}$$

$$n_0 x_2'' = \frac{\partial n_2}{\partial x} = C_1\frac{\partial\sigma_2}{\partial x} + C_2\frac{\partial\sigma_1}{\partial x}\,, \tag{10}$$

$$n_0 y_2'' = \frac{\partial n_2}{\partial y} = C_1\frac{\partial\sigma_2}{\partial y} + C_2\frac{\partial\sigma_1}{\partial y}\,. \tag{11}$$

Since within the frame of this approximation, the equations (8) - (11) determine constant second derivatives of the components of the two rays, the light beam deflections can easily be calculated.

3. APPLICATION TO THE NEIGHBOURHOOD OF CRACK TIPS

The stress model of linear elastic fracture mechanics for plane states is given by the stress components

$$\sigma_{xx} = \frac{K_I}{\sqrt{2\pi r}} \cos \frac{\Theta}{2} (1 - \sin \frac{\Theta}{2} \sin \frac{3}{2} \Theta) , \qquad (12)$$

$$\sigma_{yy} = \frac{K_I}{\sqrt{2\pi r}} \cos \frac{\Theta}{2} (1 + \sin \frac{\Theta}{2} \sin \frac{3}{2} \Theta) , \qquad (13)$$

$$\sigma_{xy} = \frac{K_I}{\sqrt{2\pi r}} \cos \frac{\Theta}{2} \sin \frac{\Theta}{2} \sin \frac{3}{2} \Theta , \qquad (14)$$

where K_I is the "mode I" stress intensity factor. The notations are explained in Figure 2. The principal stresses σ_1 and σ_2 are

$$\sigma_{1,2} = \frac{K_I}{\sqrt{2\pi r}} \cos \frac{\Theta}{2} (1 \pm \sin \frac{\Theta}{2}) , \qquad (15)$$

from which

$$\sigma_1 + \sigma_2 = \frac{K_I}{\sqrt{2\pi r}} 2 \cos \frac{\Theta}{2} , \qquad (16)$$

$$\sigma_1 - \sigma_2 = \frac{K_I}{\sqrt{2\pi r}} \sin \Theta , \qquad (17)$$

and the equivalent stress σ_v, according to the hypothesis of elastic energy distorsion,

$$\sigma_v = \sqrt{\sigma_1^2 - \sigma_1 \sigma_2 + \sigma_2^2} , \qquad (18)$$

are derived. The isochromatic fringe order of conventional photoelasticity yields

$$m = \frac{b}{S_\sigma} (\sigma_1 - \sigma_2) = \frac{b}{S_\sigma} \frac{K_I \sin \Theta}{\sqrt{2\pi r}} , \qquad (19)$$

where

$$S_\sigma = \frac{\lambda}{C_1 - C_2} . \qquad (20)$$

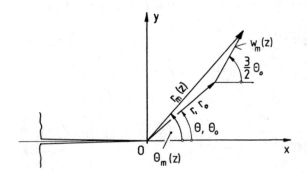

Figure 2 - Notations used for Incidence Normal to the Centre Plane of a Cracked Plate

In Figure 3 the paths of the two refracted rays are depicted. They follow from the equations (8) - (11) applied to the principal stresses (15), when a light beam is normally incident on the centre plane of the cracked plate (direction vector $\vec{s}_0$). The "isotropic" part of both rays is deflected in the plane containing $\vec{s}_0$ and forming the angle $\Theta_0 \cdot 3/2$ with the (x,z)-plane. r_0 and Θ_0 describe the point of incidence.

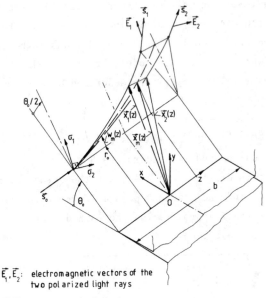

$\vec{E}_1, \vec{E}_2$: electromagnetic vectors of the two polarized light rays

$\vec{s}_1, \vec{s}_2$: wave normals of the two rays leaving the sample

Figure 3 - Propagation of the Two Refracted Rays in the Neighbourhood of a Crack Tip

The corresponding mean deflection is described by

$$\vec{w}''_m = \frac{-K_I(C_1+C_2)}{2\sqrt{2\pi}\ r_0^{3/2}\ n_0} \begin{pmatrix} \cos\frac{3}{2}\Theta_0 \\ \\ \sin\frac{3}{2}\Theta_0 \end{pmatrix}, \tag{21}$$

which yields

$$w_m = -\frac{K_I(C_1+C_2)}{4\sqrt{2\pi}\ r_0^{3/2}\ n_0}\ \frac{b_0^2}{4}\left(\left[1 + \left(\frac{2z}{b_0}\right)\right]^2 + 2B\left[1 + \left(\frac{2z}{b_0}\right)\right]\right). \tag{22}$$

The factor

$$B = -2\ \frac{\nu}{E}\ \frac{n_0-n_i}{C_1+C_2}, \tag{23}$$

considers thickness variations

$$\frac{\Delta b}{b_0} = -\frac{\nu}{E}\ (\sigma_1+\sigma_2), \tag{24}$$

where ν and E are elastic coefficients and n_i is the refractive index of the surrounding immersion (air or immersion liquid). In the cases of using matching fluid $n_i = n_0$ and of plane strain state the factor B vanishes. Moreover, (C_1+C_2) in (21) and (22) has to be replaced by $[C_1+(1+2\nu)C_2]$, when plane strain models are analyzed.

The two rays $\vec{r}_1(z)$ and $\vec{r}_2(z)$ are described by

$$\vec{w}''_{1,2} = \vec{w}''_m \pm \frac{K_I(C_1-C_2)}{8\sqrt{2\pi}\ r_0^{3/2}\ n_0} \begin{pmatrix} -3\sin 2\Theta_0 \\ \\ 1 + 3\cos 2\Theta_0 \end{pmatrix}, \tag{25}$$

what yields two curved and twisted ray paths.

Therefore it is difficult to calculate the retardation along the two rays in order to find their actual path difference for fringe order estimations. As a further approximation the path difference is taken along the mean ray, which is described by

$$r_m = r_m(z) = r_0\sqrt{1 + 2\ \frac{w_m(z)}{r_0}\ \cos\frac{\Theta_0}{2} + \frac{w_m^2(z)}{r_0^2}}, \tag{26}$$

$$\sin\Theta_m = \sin\Theta_m(z) = \frac{r_0\sin\Theta_0}{r_m}\left(1 + \frac{w_m(z)}{r_0}\ \frac{\sin\frac{3}{2}\Theta_0}{\sin\Theta_0}\right), \tag{27}$$

see Figures 2 and 3.

The isochromatic fringe order thus is approximated by

$$m = \left(1 + \frac{\Delta b}{b_0}\right) \frac{1}{\lambda} \int_{-b_0/2}^{b_0/2} \Delta n_m(r_m) \cdot \sqrt{1 + w_m'^2} \, dz \; , \tag{28}$$

where

$$\Delta n_m = n_1(r_m) - n_2(r_m) = (C_1 - C_2) \frac{K_I \sin \Theta_m}{\sqrt{2\pi} \; r_m^{3/2}} \; , \tag{29}$$

i.e., by considering equations (19), (20), (26), (27),

$$m = m_0 \left(1 + \frac{\Delta b}{b_0}\right) \frac{1}{b_0} \int_{-b_0/2}^{b_0/2} \frac{1 + \dfrac{w_m(z)}{r_0} \dfrac{\sin \frac{3}{2}\Theta_0}{\sin \Theta_0}}{\left(1 + 2\dfrac{w_m(z)}{r_0} \cos \dfrac{\Theta_0}{2} + \dfrac{w_m^2(z)}{r_0^2}\right)^{3/2}} \, dz \; , \tag{30}$$

as long as squares of the first derivatives $w_m'(z)$ can be neglected.

Equation (30) has been used to calculate fringe order errors for a crack in a stressed plate of polyester resin Palatal P, 10 mm thick, with stress intensity factor $K_I = 30 \; N/(mm)^{3/2}$. The photoelastic and elastomechanic data $C_1 = -25.5 \cdot 10^{-6} \; mm^2/N$, $C_2 = -55.5 \cdot 10^{-6} \; mm^2/N$, $(\nu/E) = 100 \cdot 10^{-6} \; mm^2/N$ and $n_0 = 1,561$ for $\lambda = 514,5 \; nm$ were taken from [6]. For convenience the calculations were carried out for loci along the y-coordinate, i.e., for $\Theta_0 = \pi/2$.

4. DISCUSSIONS AND RESULTS

Calculated results are depicted in Figure 4. The full line represents fringe order errors produced by ray path curvatures alone, while in the broken line also deflections produced by the rotation of the faces and thickness variations have been considered. Within the used stress model and the approximate mathematical analysis of light propagation, fringe order errors easily reach some ten percent.

However the diagrams of Figure 4 can only be taken as rough upper limits. This is partly due to the fact, that the assumption of a constant second derivative along a light path is not applicable for loci $r_0 < 0.5$ mm, in the given example (e.g., for $r_0 = 0.5$ mm the calculated deflection $w_m(b)$ is 0.044 mm only, but is 0.17 mm for $r_0 = 0.25$ mm). Along its path a ray passes through regions of lower refractive index gradients, and, consequently, the curvature is decreased with increasing deflection.

Therefore, in this chosen example, $r_0 = 0.5$ mm represents an upper *mathematical* limit for the applicability of the presented analysis.

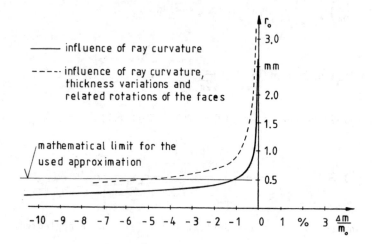

Figure 4 - Calculated Fringe Order Errors as a Function of r_0 along $\Theta_0 = \pi/2$

The application of linear elasticity seems to be less critical, since equivalent stresses, e.g., according to equation (18), are calculated to be less than the linear limit stress of 40 N/mm^2 in the case of Palatal P6 [7] for $\Theta_0 = \pi/2$ and $r_0 > 0.11$ mm.

However, further *mechanical* reasons have to be considered. It is known that stress states close to a crack tip are neither plane stress nor plane strain, but three dimensional [8]. Experiments using the shadow optical method of caustics seems to indicate, that the theory of plane stress may be applied for $r_0 > 3$ mm, when the plate is 10 mm thick, and for $r_0 > 1$ mm in the case of $b_0 = 2$ mm, respectively [9,10].

As a conclusion considerable fringe order errors produced by ray path deflections cannot be excluded in regions close to the crack tip, but the influence of other essential factors cannot yet be analyzed.

REFERENCES

1.   PINDERA, J.T., HECKER, F.W. and KRASNOWSKI, B.R., "Gradient Photo-elasticity", *Mechanics Research Communications*, Vol. 9, 1982, pp. 197-204
2.   BOKSTHEIN, M.F., "On the Resolving Power of the Polarizing System for Stress Investigations", (in Russian), *Zhurnal Tekhnicheskoi Fiziki*, Vol. 19, 1949, pp. 1103-1106.

3. PINDERA, J.T., "Technique of Photoelastic Investigations of Plane
    Stress States", (in Polish), *Rozprawy Inżynierskie*, Vol. 3, 1955,
    pp. 109-176.
4. POST, D., "Optical Analysis of Photoelastic Polariscopes", *Experimental
    Mechanics*, Vol. 10, 1970, pp. 15-23.
5. HECKER, F.W., "On the Accuracy of Photoelastic Investigations of
    Plane Stress States with High Stress Gradients", (in German),
    VDI-Berichte Nr. 480, 1983, pp. 43-49.
6. HECKER, F.W., KEPICH, T.Y. and PINDERA, J.T., "Non-Rectilinear
    Optical Effects in Photoelasticity Caused by Stress Gradients",
    Optical Methods in Mechanics of Solids, A. Lagarde, Editor, Sijthoff
    and Noordhoff, Alphen aan den Rijn, 1981, pp. 123-134.
7. PINDERA, J.T. and STRAKA, P., "On Physical Measures of Rheological
    Responses of Some Materials in Wide Ranges of Temperature and
    Spectral Frequency", *Rheological Acta*, Vol. 13, 1974, pp. 338-351.
8. PINDERA, J.T. and KRASNOWSKI, B.R., "Determination of Stress Intensity
    Factors in Thin and Thick Plates using Isodyne Photoelasticity",
    Fracture Problems and Solutions in the Energy Industry, L.A. Simpson,
    Editor, Pergamon Press, Oxford and New York, 1982, pp. 147-156.
9. SOLTÉSZ, U. and BEINERT, J., "Determination of the Stress State in
    the Surrounding of a Crack Tip by a Shadow-Optical Method", (in German),
    Report No. W6/81, Fraunhofer Institut für Werkstoffmechanik, Freiburg,
    December, 1981.
10. SHIMIZU, K., TAKAHASHI, S. and SHIMIDA, H., "Some Propositions on
    the Caustics and an Application to the Biaxial Fracture Problem",
    Proc. of the Joint Conference on Experimental Mechanics, SESA and
    ISME, Hawaii, U.S.A., 1982, pp. 201-206.

MODELLING PROBLEMS IN CRACK TIP MECHANICS   *Waterloo, Ontario, Canada*
CFC10, University of Waterloo   *August 24-26, 1983*

ON THE THREE DIMENSIONAL J INTEGRAL EVALUATION IN THE
INELASTIC AND TRANSIENT STRESS FIELDS

M. KIKUCHI, K. MACHIDA and H. MIYAMOTO
Fracture Mechanics Laboratory
Science University of Tokyo
Noda, Chiba, Japan

1. INTRODUCTION

In applying the methodology of the fracture mechanics for the reactor technology, the three dimensional evaluation of the parameter of fracture, J integral, is needed because of the complexity of the boundary conditions. Recently several papers have been published on the three dimensional crack problems in the elastic-plastic states, and the numerical methods to evaluate the three dimensional J integral have been studied [1,2]. In general, the reactor vessel is under thermal transient loadings and the cladding is deposited on the inner surface of the vessel wall. Then the general formulation of the three dimensional J integral is not available and some modifications are needed. A paper on transient loading was carried out by Aoki et al [3], and for the multi-phase problem, Kikuchi [4] extended the J integral definition.

In the following, three dimensional crack problems are analyzed using the finite element method. First, two CCT specimens with different thicknesses are analyzed and the distributions of the J integral along the crack front are obtained. The results are discussed by comparing with the results of the CT specimen. Then the semi-elliptical surface crack on the inner surface of the pressure vessel is analyzed under the thermal transient loadings. Two components of the J vector are evaluated for the curved crack problem along the crack front. The effects of the existence of the cladding are also studied and discussed.

2. ELASTIC-PLASTIC ANALYSIS OF CCT SPECIMENS

Figure 1 shows the shape, dimensions and the mesh pattern of the 1/8 portion of the CCT specimen. Two cases in which the thickness of the specimen, B, is 6 mm and 18 mm are analyzed. The mesh is divided in three layers along the specimen thickness and the J value is evaluated in the midplane of each layer along the paths shown in Figure 1 by dashed lines. Figures 2(a) and (b) show the configurations of the crack front for each specimen, which are obtained from the fatigue precracks of the experiment. The material is assumed to be SS41 steel and the material properties are shown in Table 1. The x-th component of the three dimensional J vector, J_x, is evaluated by the following equation referring to Figure 3.

$$J_x = \int_\Gamma (U\,dy - \sigma_{ij}u_{i,x}n_j dx) - \int_S (\sigma_{iz}u_{i,x})_{,z}\,dS\ , \tag{1}$$

where U is the strain energy density, σ_{ij} is the stress tensor, u_i is the displacement vector, and n_j is the unit normal on Γ.

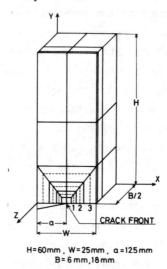

H=60mm , W=25mm , a=12.5mm
B= 6mm,18mm

Figure 1 - 1/8 Part of a CCT Specimen

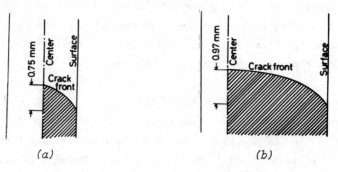

(a)   (b)

Figure 2 - Shapes of the Fatigue Precracks of CCT Specimens
(a) B = 6mm; (b) B = 18 mm

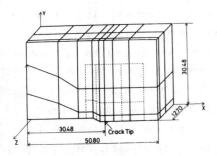

Figure 8 - 1/4 Part of the CT
Specimen

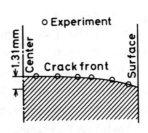

Figure 9 - Shape of the Fatigue
Precrack of 1 CT Specimen

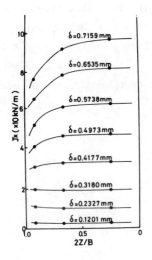

Figure 10 - J_x Distribution Along
the Crack Front

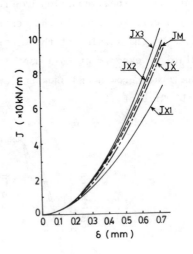

Figure 11 - J versus Displacement
Curve

3. J INTEGRAL OF A CRACK UNDER THERMAL TRANSIENT LOADINGS

The J integral of a crack in the thermal transient loading is called
$\hat{J}$ integral and is defined by Aoki et al. In case it exists in the multi-
phase material, the $\hat{J}$ integral definition is introduced by Kikuchi. Both
are expressed as follows (see Figure 12):

$$\hat{J}_\ell = \int_{\Gamma_1} (U^e \delta_{j\ell} - \sigma_{ij} u_i,_\ell) ds_j + \int_{A_1} \sigma_{ij} \varepsilon^*_{ij},_\ell dA \ , \tag{2}$$

$$\hat{J}_\ell = \int_{\Gamma_1-\Gamma_2} (U^e \delta_{j\ell} - \sigma_{ij} u_i,_\ell) ds_j + \int_{A_1-A_2} \sigma_{ij} \varepsilon^*_{ij},_\ell dA \ , \tag{3}$$

Though the J_x value is the maximum nearby the free surface, the stretched zone width is the maximum at the centre of the specimen. It is because the J_{IC} criterion is not available nearby the free surface due to the plane stress condition. Experimentally the J integral of the CCT specimen is evaluated by using the conventional equation proposed by Rice et al [5] from P-δ curves, and is denoted by J_R. The average value of the J_x values over the crack front can be also obtained from Figure 5 and is denoted by J_x'. Figures 7(a) and (b) show each J value for two specimens. It is noticed that the J_R value is very much similar to J_x' value for both specimens, and they are comparatively similar to the J_{x3} value, which is the J_x value at the centre of the specimen. That is, though the J values change largely along the crack front, the average value can be evaluated by the conventional equation by Rice et al. The fact that crack growth mainly occurs at first in the centre of the plate and that J_R value is similar to the J value at the centre of the plate mean that the J value obtained by experiment is a useful parameter to describe the condition of the crack growth and show the availability of the conventional equation by Rice et al.

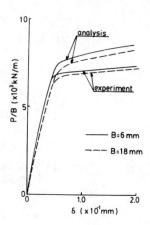

Figure 4 - Load versus Displacement Curves

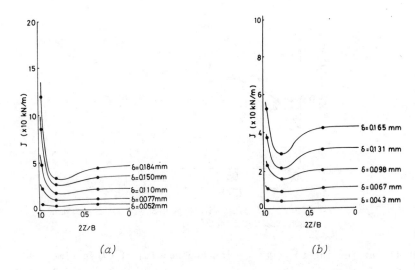

(a)   *(b)*

Figure 5 - J_x Distributions Along the Crack Front
(a) B = 6mm; (b) B = 18 mm

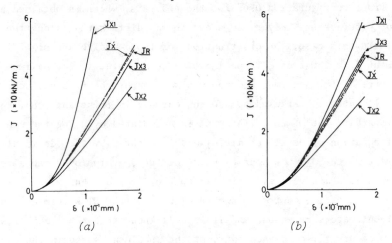

(a)   *(b)*

Figure 6 - J versus Displacement Curves
(a) B = 6mm; (b) B = 18 mm

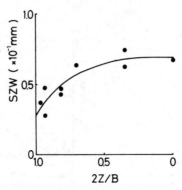

Figure 7 - Stretched Zone Width for B = 6 mm Specimen
Obtained by Experiment

It is interesting to compare with the results of CT specimen which has been analyzed by authors and presented previously. Figure 8 shows the dimensions and mesh pattern of the 1 CT specimen. Figure 9 is the crack front configuration used for the analysis, which is obtained by the fatigue precracking. Figure 10 shows the J_x distribution along the crack front. In this case, the distributions of the J_x value are significantly different from those of CCT specimens. The J_x value is the maximum in the centre of the specimen and the minimum nearby the free surface. Figure 11 shows J_x versus displacement curve. In these analyses, the experimental value of the J integral is calculated by using the conventional equation by Merkle et al [6] and is denoted by J_M. It is also noticed that the J'_x, the average value of the J_x value over the specimen thickness is again very much similar to the J_M value, and also similar to the J_x value at the centre of the specimen. On the CT specimen, many experiments are carried out and it is well known that the crack growth usually occurs first in the centre of the specimen, so again the conventional method to evaluate the J value is useful for the condition of the crack growth.

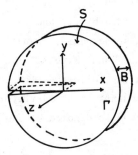

*Figure 3 - Integral Path and Surface of the Three-Dimensional
        J Integral*

Table 1 - Material Properties of CCT Specimen

Young's Modulus E(GPa)	206
Poisson's Ratio ν	0.3
Yield Stress σ_Y (MPa)	265.0
Constitutive Equation	$\bar{\sigma} = 775(0.015 + \bar{\varepsilon}_p)^{0.26}$ (MPa)

Figure 4 shows the load P versus displacement δ curves obtained by
the analyses. As the displacement method is used in the finite element
analysis, the curves by analyses are a little higher than those by experiments and the tangentials in elastic part agrees very well. Figures 5(a)
and (b) shows the distributions of the J_x value along the crack front in
the elastic-plastic deformation. When δ is small, the J value are nearly
uniform along the crack front but as the deformation proceeds, the distributions change largely. The J_x value near the free surface becomes
significantly larger than that at the centre of the specimen in both
cases. For B = 6 mm specimen, the ratio of the J_x near the surface to
that in the centre is nearly 2.5, and for B = 18 mm specimen it is nearly
1.3. The fact that the ratio of thinner specimen is larger than that of
the thick specimen suggests that the change of the stress states, from the
plane stress condition nearby surface to the plane strain in the centre
occurs abruptly in the thin specimen. In general, it is considered that
the crack growth occurs first where the J value exceeds the critical J
value. Figure 6 shows one example of the distribution of the stretched
zone along the crack front for B = 6 mm specimen obtained by experiments.

where U^e is the elastic strain energy density, ε^*_{ij} is the eigen strain
tensor, A_1 is the area surrounded by Γ_1, and Γ_2, A_2 means the contour and
area corresponding to the phase boundary. Equations (2) and (3) can easil
be extended to the three dimensional stress field as in a similar manner
to that of equation (1). In the following, the crack front has some
curvature, then the two components of the $\hat{J}$ vector, $\hat{J}_x$ and $\hat{J}_y$ are evaluate
and the magnitude of the vector is denoted by $\overline{\hat{J}}$.

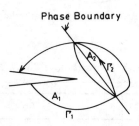

Figure 12 - Contour for the J Integral

    Figure 13 shows the model of the surface crack on the inner surface
of a pressure vessel. The aspect ratio, a/c is 0.8. The thermal transien
loading is applied to the model due to the injection of the coolant in
the vessel as shown in Figure 13. The material constants for the analyses
are shown in Table 2. The change of the $\overline{\hat{J}}$ values along the crack front
with the time lapse is shown in Figure 14. It is noted that the $\hat{J}$ value
increases rapidly at first, but after some time lapse, it begins to
decrease. These are due to the contraction of the body by the decrease
of the temperature of the vessel wall. These phenomena are preferable
from the viewpoint of the safety design of the pressure vessel.

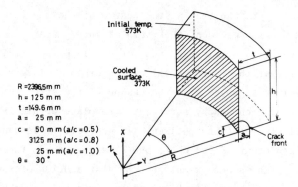

Figure 13 - Model of a Surface Crack of a Pressure Vessel

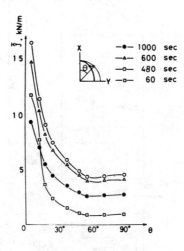

Figure 14 - Change of the $\bar{J}$ Value with Time Lapse

Table 2 - Material Constants for the Thermal Transient Analysis

	Base Metal	Clad
Young's Modulus (GPa)	205.94	195.94
Poisson's Ratio	0.3	0.3
Coefficient of Thermal Expansion (1/°K)	4.027×10^{-8}	5.858×10^{-8}
Thermal Conductivity (W/m°K)	41.53	16.58
Thermal Diffusivity (m²/s)	1.056×10^{-5}	4.139×10^{-5}
Heat Transfer Coefficient (W/m²)	1.136×10^3	1.136×10^3

Figure 15 shows the crescent type internal crack under the cladding, which are observed frequently due to the stress concentration by the depositing process of the cladding. The three dimensional $\hat{J}_\ell$ integral is analyzed by using equation (3) around the crack front. The results are shown in Figures 16(a) and (b). The $\hat{J}$ value at the straight portion of the crack front is larger than that of the circular front. Figure 17 shows the comparison of the $\hat{J}$ values with those by two dimensional model analyses at both points A and B in the figure. It is noticed that the two and three dimensional analyses agree quantitatively for both points and it is sufficient by the two dimensional analyses in this case.

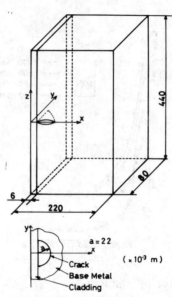

Figure 15 - *Model of a Three-Dimensional Underclad Crack*

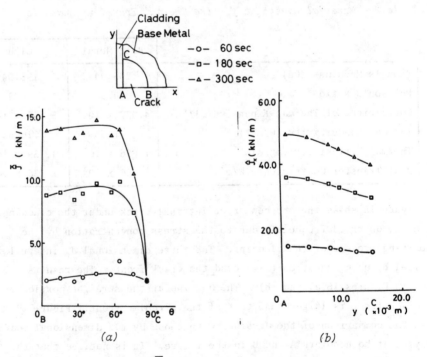

Figure 16 - *Change of the $\overline{\widehat{J}}$ Values Around the Underclad Crack Front*
(a) $\overline{\widehat{J}}$ Values along B-C; (b) $\overline{\widehat{J}}$ Values along A-C

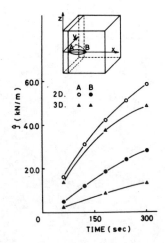

Figure 17 - Comparison with the Results by Two Dimensional Analysis

4. CONCLUDING REMARKS

Several three dimensional crack problems are treated and it is shown that the fracture mechanics methodology is applicable to the reactor vessel design by using the method shown in this paper. Due to the limitation of the pages, only a small portion of the results are shown. Details will be published elsewhere.

REFERENCES

1. KIKUCHI, M., et al, *Trans. of the 5th SMiRT*, G7/2, 1979.
2. BAKKER, A., *Trans. of the 6th SMiRT*, G3/3, 1981.
3. AOKI, et al, *J. of Applied Mechanics*, Vol. 48, 1981, p. 825.
4. KIKUCHI, M., et al, *Numerical Methods in Fracture Mechanics*, D.R.J. Owen, Editor, 1980, p. 295.
5. RICE, J.D., et al, *ASTM STP 536*, 1973, p. 321.
6. MERKLE, J.G., et al, *Trans. of the ASME, Series J*, Vol. 96, 1974, p. 286.

MODELLING PROBLEMS IN CRACK TIP MECHANICS   *Waterloo, Ontario, Canada*
CFC10, University of Waterloo   *August 24-26, 1983*

ANALYSIS OF MODELS OF STRESS STATES IN THE REGIONS OF
CRACKS USING ISODYNE PHOTOELASTICITY

J.T. PINDERA and B.R. KRASNOWSKI   M.-J. PINDERA
Institute for Experimental Mechanics   CertainTeed Corporation
University of Waterloo   Blue Bell, Pennsylvania
Waterloo, Ontario, Canada   U.S.A.

1. INTRODUCTION

The fact that the general and engineering solutions of plane elasticity yield only approximate information on stress fields in regions of crack tips and notches has been known in engineering from more than half a century [1]. Strongly nonhomogeneous deformation fields must lead to clearly three dimensional local stress fields. This is true for engineering structures such as plates of finite thickness, regardless of the aspect ratio. However, due to analytical and experimental difficulties associated with determining such fields, only a limited number of three-dimensional solutions to these problems exist [2,3,4,5,6]. The majority of present-day closed form solutions is based on such simplifying assumptions as the existence of plane strain or stress states which, in practice, are seldom approximated.

Since it is not possible to assess analytically the reliability and accuracy of models of such stress states because the most general theory does not exist, it is necessary to verify experimentally their predictive power.

In this paper it is shown that the newly developed nondestructive methods of isodyne photoelasticity are readily applicable for measuring the pronounceably spatially distributed actual stress states in the regions of geometric and material discontinuities such as cracks, notches, joints and material interfaces. These methods are self-contained and yield values of stress components without any auxiliary relations. The underlying concept of a particular type of isodynes - called elastic isodynes - which is based on the two-dimensional classical theory of linear elasticity, is presented in Figure 1. The experimentally obtained photoelastic isodynes represent a generalization of plane elastic isodynes, and carry information on all components of the stress tensor [7-12].

ISODYNES OF PLANE STRESS STATES
(A SAMPLE OF SIMPLEST CASES)

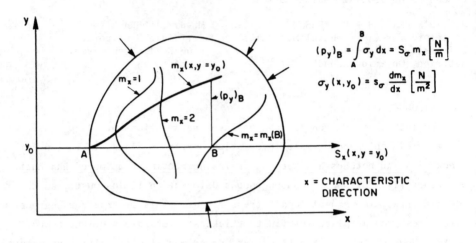

$$(p_y)_B = \int_A^B \sigma_y \, dx = S_\sigma \, m_x \left[\frac{N}{m}\right]$$

$$\sigma_y (x, y_0) = s_\sigma \frac{dm_x}{dx} \left[\frac{N}{m^2}\right]$$

Figure 1 - Concept of Elastic Isodynes (a family of characteristic lines of plane stress fields)

Techniques of isodyne photoelasticity directly yield values of normal and shear stress components and principal directions in chosen cross-sections in chosen planes within the object of investigation. A particular technique of isodyne coatings supplies values of normal stress components at the surface of the object. Furthermore, it ought to be mentioned that the theory of isodyne photoelasticity allows to determine regions in stressed bodies where linear elasticity theories are not applicable.

The presently developed technique of isodyne photoelasticity require that the object or a reduced model of it be transparent in the visible or infrared band of electromagnetic radiation.

Two standard isodyne techniques have been applied to obtain data on three-dimensional stress states in regions of crack tips, presented in this paper: the technique based on plane isodynes (related to elastic isodynes) and the technique based on generalized isodynes.

2. CONCEPTUAL PROBLEMS REGARDING APPLICABILITY OF THE LINEAR ELASTIC FRACTURE MECHANICS IN LIGHT OF EXPERIMENTAL RESULTS

It had been shown [7-12] that the isodyne techniques satisfy the two basic requirements of any theory: the conclusions drawn are self-consistent (noncontradictory with the assumptions), and the ranges of applicability of the assumptions are empirically determinable. Thus, the empirical results such as those shown in Figures 2 and 3 can be taken as measures for the applicability of the analytical solutions for stress states in regions of cracks - this applied to the linear solutions, in particular, with all the consequences regarding the definition of K_I-factor; it should be noted that the character of variation of values of stress components with the distance from the middle plane of the beam does not depend on the height-to-thickness ratio. Because of this dependence, the values of stress components normal to the radial direction from the crack tip are not unique.

The theoretical value of $K_I = K_{th}$ was based on Benthem and Koiter asymptotic solution [15] for an infinite strip of the width h, with a single edge crack of the length a, loaded by a bending moment M, which is presented in the form of the asymptotic expansion of the stress singularity with regard to $\frac{a}{h}$,

$$\sigma_{xx} \rightarrow 1.122 \frac{6M}{(h-a)^{3/2}\sqrt{2r}} (1 - 1.217 \frac{a}{h}) (\frac{a}{h})^{1/2} [1 + 0 (\frac{a}{h})^2] ,$$

where $0 (\frac{a}{h})^2$ denotes a quantity of the order of $(\frac{a}{h})^2$.

The stress intensity factor therefore has the form:

$$(K_I)_{th} = \lim_{r \to \infty} \sigma_{xx} \sqrt{2\pi r}$$

$$= 1.22\sqrt{\pi} \frac{6M}{(h-a)^{3/2}} (1 - 1.217 \frac{a}{h})(\frac{a}{h})^{1/2} \cdot [1 + 0 (\frac{a}{h})^2] ,$$

or

$$(K_I)_{th} = \frac{6M}{(h-a)^{3/2}} f (\frac{a}{h}) ,$$

where

$$f (\frac{a}{h}) = 1.122\sqrt{\pi} (\frac{a}{h})^{1/2} (1 - 1.217 \frac{a}{h}) [1 + 0 (\frac{a}{h})^2] .$$

STRESS INTENSITY FACTOR K_I IN BEAMS WITH CRACK IN BENDING

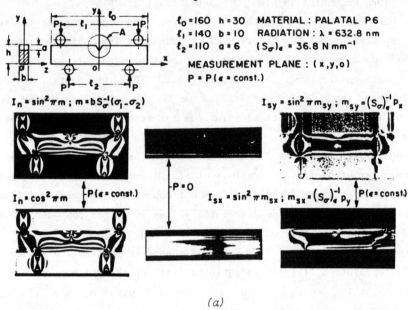

$l_0 = 160$ $h = 30$ MATERIAL : PALATAL P6

$l_1 = 140$ $b = 10$ RADIATION : $\lambda = 632.8$ nm

$l_2 = 110$ $a = 6$ $(S_\sigma)_\epsilon = 36.8$ N mm^{-1}

MEASUREMENT PLANE : (x,y,o)

$P = P(\epsilon = \text{const.})$

$I_n = \sin^2 \pi m$; $m = bS_\sigma^{-1}(\sigma_1 - \sigma_2)$

$I_{sy} = \sin^2 \pi m_{sy}$; $m_{sy} = (S_\sigma)_\epsilon^{-1} p_x$

$I_n = \cos^2 \pi m$ $P(\epsilon = \text{const.})$

$P = 0$

$I_{sx} = \sin^2 \pi m_{sx}$; $m_{sx} = (S_\sigma)_\epsilon^{-1} p_y$ $P(\epsilon = \text{const.})$

(a)

Figure 2 – *Prismatic Rectangular Beams with Cracks of the Depth "a",*
in Pure Bending. Diagrams of normal stresses at varying
distances from the crack tip, and of the stress intensity
functions $K_I(r)$ determined experimentally, using photoelastic
isodyne techniques, for the middle planes of the beams.
Theoretical values of the stress intensity factors K_I are
denoted by heavy points.

(a) Crack in Tension; $a = 6$ mm, $(K_I)_a$;
(b) Crack in Compression; $a = 3$ mm, $(K_I)_b$;
(c) Results for (a) and (b).

(continued)

(continued)

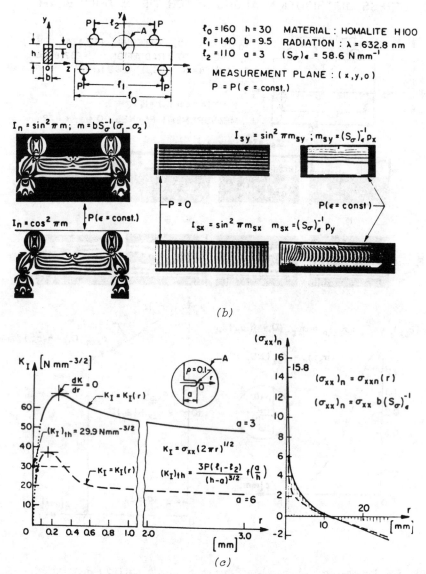

(b)

(c)

Figure 2 – Prismatic Rectangular Beams with Cracks of the Depth "a",
in Pure Bending. Diagrams of normal stresses at varying
distances from the crack tip, and of the stress intensity
functions $K_I(r)$ determined experimentally, using photoelastic
isodyne techniques, for the middle planes of the beams.
Theoretical values of the stress intensity factors K_I are
denoted by heavy points.

(a) Crack in Tension; $a = 6$ mm, $(K_I)_a$;
(b) Crack in Compression; $a = 3$ mm, $(K_I)_b$;
(c) Results for (a) and (b).

3D-STRESS DISTRIBUTION ALONG NOTCH TIP IN THICK BEAM

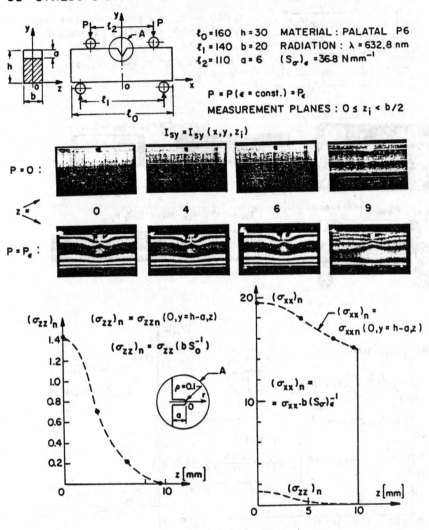

Figure 3 - *Prismatic Rectangular Beam with Crack in Pure Bending.*
Components of the three-dimensional stress state along the
notch tip, determined by means of isodyne photoelasticity:

$$\sigma_{zz} = \sigma_{zz}(x = 0,\ y = h-a,\ z),$$
$$\sigma_{xx} = \sigma_{xx}(x = 0,\ y = h-a,\ z),$$
$$\sigma_{yy} = 0.$$

Since (a/h) is one of the order of 1/10 the quadratic term involving parameter (a/h) can be neglected and

$$f\left(\frac{a}{h}\right) \equiv 1.122\sqrt{\pi}\left(\frac{a}{h}\right)^{1/2}\left(1 - 1.217\frac{a}{h}\right).$$

The above equation has been used to calculate the theoretical values of the stress intensity factor $(K_I)_{th}$ shown in Figure 2.

Considering the fact that the stress components normal to any chosen direction, as well as the regions of nonlinear local material behaviour of the crack tip can be readily determined by using the isodyne techniques, it is proposed to introduce an empirical function, limit value of which defines the plane stress or plane strain intensity factor K_I:

$$K_I(r,\theta) = \sigma_{nr}(r,\theta)\ (2\pi r)^{1/2},$$

when $\theta = 0$

$$K_I(r) = [\sigma_{nn}(r)]\ (2\pi r)^{1/2},$$

and to call it the stress intensity function, in analogy with the terminology used in some plasticity theories. In the above expression, the stress component σ_{nn} normal to the crack direction r is a function of the distance from the crack tip r, and also of the distance from the middle plane of the plate z_o too:

$$\sigma_{nn} = \sigma_{nn}\ (r,\ z_o).$$

It is easy to generalize this function for an arbitrary direction of r.

The $K_I(r)$ function represents the actual stress field at the crack tip and can be used as a measure to assess various analytical-experimental methods used to determine the values of the classical stress intensity factor K_I.

3. ANALYSIS OF LAMINATED STRUCTURES WITH MATERIAL AND GEOMETRIC DISCONTINUITIES

The major task of modern experimental mechanics is to provide empirical bases for the development and/or verification of the assumptions and simplifications employed in the construction of the analytical solutions. This is of a particular importance in newly developing fields such as the

theory and technology of composite structures. Consequently, the method
of isodynes has been used to collect information on stress states in
laminated structures, which are difficult to obtain using other experimental
methods.

Figure 4 presents a sample of experimental and analytical results
associated with the longitudinal tension of three-ply polyester/aluminum
and polyester/glass laminated structures with a transverse crack in the
middle ply.

The sensitivity of the isodyne photoelasticity to the disturbances of
the stress fields caused by cracks is clearly illustrated in Figure 4(a),
at various elevations and various distances from the plane of lamination.

Figures 4(b) and 4(d) illustrate the distribution of the experimentally
determined stresses which are caused by the crack in two laminated structures
having different ratios of the material parameters E and ν. The influence
of the material parameters, particularly that of the Poisson ratios, is
clearly visible; specifically the loci of the maximum shear stress σ_{13}
appear to form straight lines with radically different inclinations with
respect to the plane of the crack.

The correlation between the empirical results and the predictions of
the approximate analytical model of reference [16], in which the variation of the normal axial stress with the ply thickness is neglected, is
mainly of a qualitative nature at points sufficiently remote from the
crack; the character of the empirical and analytical stress functions
differs radically in the region close to the crack; whereas the analytical
solution yields values which increase continuously, the empirical results
indicate local maxima at small distances from the crack.

On the other hand, a model which is more rigorous with respect to the
variation of stresses with the distance from crack, presented by Zak and
Williams [17,18], in the form

$$\sigma_{ij}(r,\theta) = r^{\lambda-1} \, F_{ij}(\theta) \ ,$$

does predict qualitatively similar stress distribution in the vicinity of
the crack tip, with regard to the existence of σ_{13}-maxima, Figures 4(c),
4(e) and 4(f). Comparing the predictions of the plane stress model with
the prediction of the plane strain model one can notice that the character
of the curves is similar with respect to location of maxima; however, the

plane stress model yields higher values of the normalized maximum shear
stresses then the plane strain model does.

ISODYNE FIELDS IN OUTER PLIES OF A THREE-PLY
STRUCTURE WITH INTERNAL CRACK

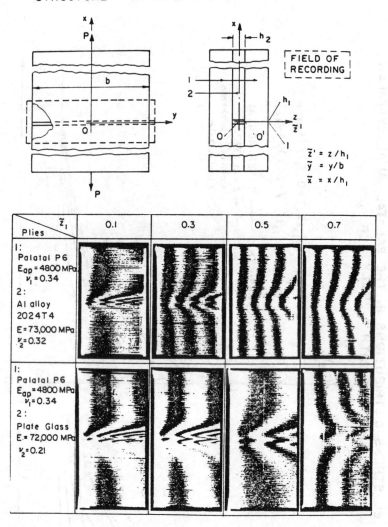

Figure 4 - *Three-ply Structures with a Cracked Internal Ply, in Tension.*
Diagrams of normalized shear stresses in the plane of symmetry
normal to the face of the structure determined analytically and
experimentally, for two structures: Polyester-Al. Alloy-
Polyester (P-Al.Al.-P), and Polyester-Glass-Polyester, (P-G-P).

(a) Scheme of the System and Recordings of Isodynes

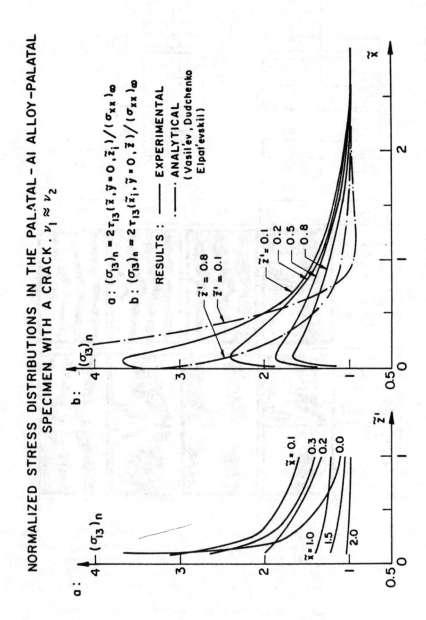

Figure 4(b) – Empirical and Analytical Data for (P-Al.Al-P)-Specimen

NORMALIZED STRESS DISTRIBUTION IN THE
PALATAL-AI ALLOY-PALATAL SPECIMEN WITH CRACK

ANALYTICAL RESULTS (Zak , Williams)
PLANE STRAIN MODEL $\nu_1 \approx \nu_2$

$$(\sigma_{13})_n = \sigma_{13}/d_2$$

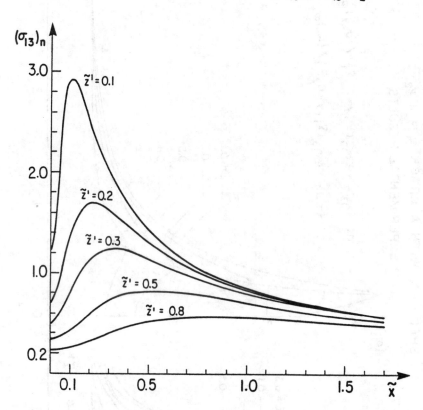

Figure 4(c) - Analytical Data for (P-Al.Al-P)-Specimen According to Plane Strain Model

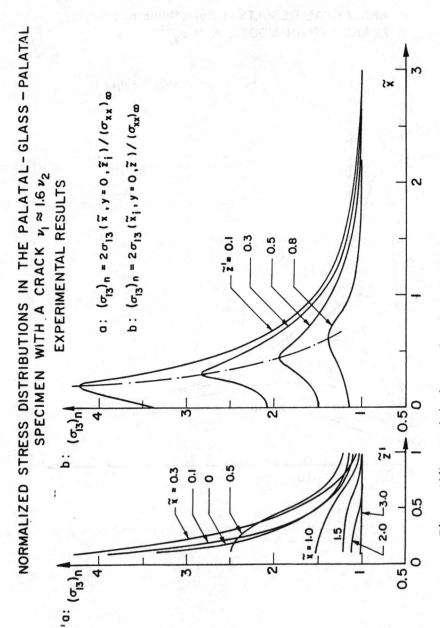

Figure 4(d) – Empirical Data for (P-G-P)-Specimen

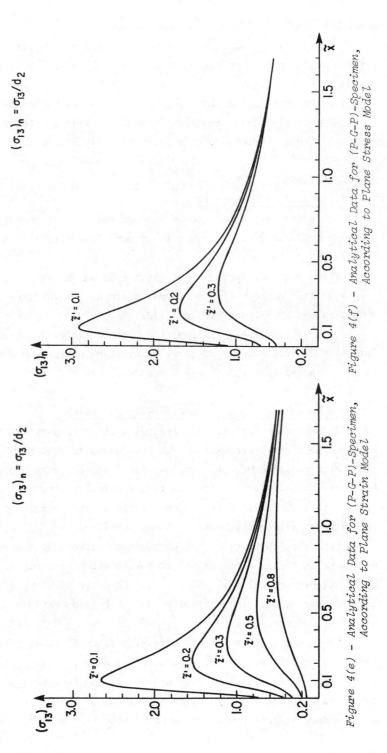

NORMALIZED STRESS DISTRIBUTION IN THE
PALATAL-GLASS - PALATAL SPECIMEN WITH CRACK

ANALYTICAL RESULTS (Zak, Williams)
PLANE STRESS MODEL $\nu_1 \approx 1.6\,\nu_2$

$(\sigma_{13})_n = \sigma_{13}/d_2$

*Figure 4(f) – Analytical Data for (P-G-P)-Specimen,
According to Plane Stress Model*

NORMALIZED STRESS DISTRIBUTION IN THE
PALATAL-GLASS - PALATAL SPECIMEN WITH CRACK

ANALYTICAL RESULTS (Zak, Williams)
PLANE STRAIN MODEL $\nu_1 \approx 1.6\,\nu_2$

$(\sigma_{13})_n = \sigma_{13}/d_2$

*Figure 4(e) – Analytical Data for (P-G-P)-Specimen,
According to Plane Strain Model*

4. OTHER PERTINENT FEATURES OF ISODYNE METHODS

The reliability and resolution of experimental results is sufficiently high to allow them to be taken as a measure for accuracy of the analytical models of stress states.

It ought to be noted that the same isodyne method can be used to determine the technologically important thermal as well as the thermally induced residual stresses which are of a particular importance in fracture mechanic

5. CONCLUSIONS

5.1 Efficacy of the New Method

- There does not appear to be another known experimental method which supplies nondestructively such sets of empirical data within comparable constraints.

- There does not appear to be another known empirical method based on such a limited number of assumptions regarding the involved physical phenomena and the stress/strain fields.

- This is a self-verifying method: the empirically obtained light intensity distributions also carry information whether or not the conditions following from the theory of isodynes are satisfied.

5.2 Some Modelling Problems in Crack-Tip Mechanics

Obtained empirical evidence regarding the actual stress fields in the neighbourhood of the crack leads to the following conclusions:

- Extrapolation of the solutions based on the concept of plane stress intensity factor (derived for the infinitesimally thin or infinitely thick plates, that is for infinite space) for the actual stress states existing in plates of finite thicknesses is not justified.

- For practical engineering purposes the stress concentration caused by cracks in the homogeneous and composite structures can be represented by solutions containing singularities, the order of which depends on the material parameters (in particular, on the Poisson ratios) and on some geometric parameters.

- The concept of the stress intensity function leads to a more representative description of the stress fields near cracks and facilitates the assessment of the reliability of analytical and numerical solutions. In particular, this concept supplies a rational basis for techniques which utilize some singular solutions supplemented by some experimentally

evaluated data on the stress values (e.g., on $\sigma_1 - \sigma_2$) at points sufficiently remote from the crack tip which are obtained for real stress
fields not containing singularities.

- It appears that the isodyne photoelasticity, yielding directly
values of normal and shear stress components along arbitrarily chosen
directions, can supply data on the actual stress fields, which are needed
to assess the influence of the singular term in the analytical solution
on the predicted stress values.

6. ACKNOWLEDGEMENT

Research had been supported by the National Sciences and Engineering
Research Council of Canada under Grant A-2939.

REFERENCES

1. THUM, A. and SVENSON, O., "Die Verformungs-und Beanspruchungsverhaltnisse von glatten und gekerbten Staben, Scheiben und Platten in Abhangigkeit von deren Dicke und Belastungsart", *Forschung Ing.-Wes.*, No. 11, 1942.
2. THUM, A, PETERSEN, C. and SVENSON, O., *Verformung, Spannung und Kerbwirkung, (Deformation, Stress and Notch Effect)"*, VDI-Verlag, Düsseldorf, 1960.
3. FROCHT, M.M. and LEVEN, M.M., "On the State of Stress in Thick Bars", *J. of Applied Physics*, Vol. 13, 1942, pp. 308-313.
4. STERNBERG, E. and SADOVSKY, M.A., "Three-Dimensional Solution for the Stress Concentration Around a Circular Hole in a Plate of Arbitrary Thickness", *J. of Applied Physics*, Vol. 20, 1949, pp. 27-38.
5. ELLYIN, F., LIND, N.C. and SHERBOURNE, A.N., "Elastic Stress Field in a Plate with a Skew Hole", *J. Eng. Mech. Div., ASCE*, Vol. 92, No. EM1, 1966, pp. 1-10.
6. SIH, G.C., "A Review of the Three-Dimensional Stress Problem for a Cracked Plate", *Int. J. of Fracture Mechanics*, Vol. 7, 1971, pp. 39-61.
7. PINDERA, J.T. and KRASNOWSKI, B.R., "Determination of Stress Intensity Factors in Thin and Thick Plates Using Isodyne Photoelasticity", *Proc. of the Fifth Canadian Fracture Conference*, edited by L.A. Simpson, Springer-Verlag, 1982.
8. PINDERA, J.T., "Analytical Foundations of the Isodyne Photoelasticity", *Mechanics Research Communications*, Vol. 8, 1981, pp. 391-397.
9. PINDERA, J.T. and MAZURKIEWICZ, S.B., "Photoelastic Isodynes: A New Type of Stress-Modulated Light Intensity Distribution", *Mechanics Research Communications*, Vol. 4, 1977, pp. 247-252.
10. PINDERA, J.T., KRASNOWSKI, B.R. and PINDERA, M.-J., "Determination of the Interface Stresses in Composite Materials", *Proc. of the 1982 Joint Conference (JSME/SESA)*, Hawaii, 1982, pp. 18-22.
11. PINDERA, J.T., KRASNOWSKI, B.R. and PINDERA, M.-J., "An Analysis of Semi-Plane Stress States in Fracture Mechanics and Composite Structures Using Isodyne Photoelasticity", *Proc. of the 1982 Joint Conference (JSME/SESA)*, Hawaii, 1982, pp. 417-421.

12. PINDERA, J.T. and KRASNOWSKI, B.R., "IEM-Paper No. 1, Theory of Elast
     and Photoelastic Isodynes", *SM Paper No. 184*, Solid Mechanics Divisio
     University of Waterloo, Ontario, Canada, October, 1983.
13. PINDERA, J.T., HECKER, F.W. and KRASNOWSKI, B.R., "New Experimental
     Method: Gradient Photoelasticity", *Proc. 8th Canadian Congress of
     Applied Mechanics*, Moncton, Universite de Moncton, 1981, pp. 517-518.
14. PINDERA, J.T., HECKER, F.W. and KRASNOWSKI, B.R., "Gradient Photo-
     elasticity", *Mechanics Research Communications*, Vol. 9, No. 3, 1982,
     pp. 197-204.
15. BENTHEM, J.P. and KOITER, W.T., "Asymptotic Approximation to Crack
     Problems", *Mechanics of Fracture*, Vol. 1, edited by G.C. Sih, Noord-
     hoff International Publishing, Leyden, 1973, pp. 131-177.
16. VASIL'EV, V.V., DUDCHENKO, A.A. and ELPAT'EVSKII, A.N., "Analysis
     of the Tensile Deformation of Glass-Reinforced Plastics", *Mekhanika
     Polimerov*, No. 1, January-February, 1970, pp. 144-147.
17. ZAK, A.R. and WILLIAMS, M.L., "Crack Point Stress Singularities at
     a Bi-Material Interface", *J. Appl. Mech.*, Vol. 30, 1963, pp. 142-143.
18. ZAK, A.R. and WILLIAMS, M.L., "Crack Point Stress Singularities at
     a Bi-Material Interface", *GALCIT SM 62-1*, California Institute of
     Technology, Pasadena, January, 1962.

MODELLING PROBLEMS IN CRACK TIP MECHANICS   *Waterloo, Ontario, Canada*
CFC10, University of Waterloo   *August 24-26, 1983*

A CALIBRATION TECHNIQUE FOR THE D.C. POTENTIAL DROP METHOD OF
MONITORING THREE DIMENSIONAL CRACK GROWTH

D.J. BURNS, R.J. PICK and D. ROMILLY
Department of Mechanical Engineering
University of Waterloo
Waterloo, Ontario, Canada

1. INTRODUCTION

One of the major problems in the experimental study of fracture is the
accurate determination of defect size, shape and growth in metallic specimens or components. Various NDT techniques exist for identifying defects
but only ultrasonic techniques have been applied to accurately sizing
defects and monitoring their growth. Recently Potential Drop techniques
have also become available for measuring the one dimensional growth of
defects in two dimensional specimens. If one wishes to measure the two
dimensional growth of a defect in a three dimensional body such as a surface
crack, then the techniques of measurement are not well established. The
authors have developed Potential Drop techniques for the measurement of
growth of long shallow defects in line pipe girth welds and for elliptical
surface cracks in tensile specimens. For long shallow defects, multiple
probes along the defect have proven to be effective in determining the
defect profile as it grows in fatigue or ductile tearing. Such detail is
often necessary since defect growth may not be uniform.

A basic difficulty of the Potential Drop technique is one of calibration.
Normally a series of calibration tests are undertaken with defects of known
dimensions. Such techniques are convenient for one dimensional defect
growth however require considerable effort for two dimensional defects
which can grow in both length and depth. This is particularly true if
multiple probes are in use. The authors have developed an electrolyte bath
analogy suggested by Knott [1] to aid in the calibration of flat plates
with either one or two dimensional cracks. This paper describes the analogue
system and compares the calibration results with experiments. Several
calibration curves generated with the electrolyte bath analogy are also
included.

2.   POTENTIAL DROP EQUIPMENT

At present, two D.C. crack monitoring systems are in use at the Univer-sity of Waterloo. The larger system, designed to monitor crack growth in large diameter line pipe, consists of a 50A constant current supply with a single channel differential amplifier. Multiple probes are monitored through the use of a low voltage switch box preceding the amplifier. The second system employs a 25A constant current supply to be used on smaller specimens and six differential amplifiers for use with six probes. The amplifiers may be used interchangeably with the power supplies depending on the current level required for the application. In both systems, a tracking differential amplifier sends a timing pulse to the power supply which outputs a preset pulsed D.C. current. The amplifier tracks the pulse and samples the probe voltages at both a stable maximum current and at zero current. Both amplifiers are equipped with a gain of 200 to 2000 which is selected according to the magnitude of the probe output signal. A common-mode rejection system is also employed to eliminate noise, and it is, therefore, necessary to attach three leads to the specimen surface. This is done by soldering the voltage leads to tabs which have been spot-welded to the specimen surface. Normally, the original uncracked voltage or the voltage from a set of probes located in the current field remote from the crack is used as a reference Potential Drop. This reference is used to normalize the output voltage from the probes across the crack.

3.   MONITORING CRACK GROWTH IN LARGE DIAMETER PIPE

In order to investigate the behaviour of surface cracks found in the girth welds of large diameter pipelines, full scale pipe fracture tests were undertaken at both the University of Waterloo and the Welding Institut of Canada [2,3]. In several of these tests, stable crack growth was sus-pected of occurring prior to final failure of the weld. In order to mea-sure this crack growth and to assist in the preparation of initial fatigue cracks, the DC Potential Drop system was used. Two specimens were prepared containing long shallow internal fatigue cracks in the girth weld joining lengths of 914 mm diameter line pipe.

A Potential Drop calibration curve (Figure 1) was developed using both saw cuts and fatigue cracks grown in similar pipe. On the test pipe, pairs of probes were located 1.6 mm from the crack plane along the length of the defect (Figure 2). The current (40A) was input to the pipe using a point

contact 102 mm from the crack plane. It was noted that since a point
source was used in essentially an infinite plate, the sensitivity of the
Potential Drop was a function of the distance from the line joining the
current input points.

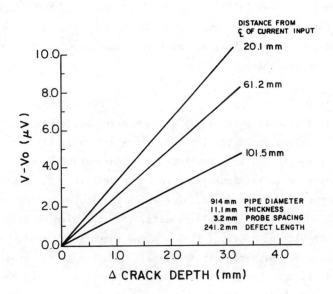

Figure 1 - *Potential Drop Calibration Curves for Large Diameter Pipes*

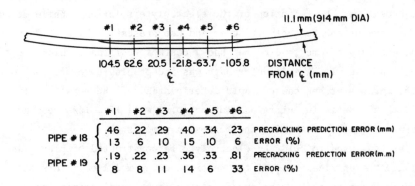

Figure 2 - *Potential Drop Probe Locations and Results for Large
Diameter Pipe Precracking*

To develop a long shallow defect, a fatigue crack was grown from a series of slitting saw cuts using single point loading at a series of sites along the defect. The crack depth was monitored with the Potential Drop probes and the error between the predicted depth and actual depth for the specimens is shown in Figure 2. Stable crack growth was also measured using the Potential Drop probes during the fracture tests. Full fracture occurred in both tests making it impossible to determine the error in the Potential Drop readings.

4. CALIBRATION

The most difficult part of any crack monitoring system is its calibration. Since only a few analytical solutions are available for one-dimensional crack growth and less for two-dimensional crack growth, most calibrations make use of measurements made on known defects such as saw cuts, milled grooves and fatigued specimens which are broken or cut open. While these techniques give reasonable results, they can also be very time consuming, difficult and costly. This led the authors to develop an analogue calibration method suggested by Knott [1]. This technique uses an electrolyte bath in the shape of the component (Figure 3) as the continuum while models of the defect made of a nonconducting material are inserted into the electrolyte to represent the defect. The voltage drop anywhere across the crack plane can then be measured by using a set of insulated probes inserted into the electrolyte. The most difficult part of the development was the selection of the electrolyte solution. While mercury is the obvious choice, both its cost and use present problems. Powdered metal and carbon were studied but had a contact resistance which was too high to be useful. A salt-water bath was acceptable, however, the corrosion of the various probes caused rapid deterioration of the results. It was found that a potassium nitrate solution works well if a low frequency A.C. current is used to stabilize the corrosion process.

The tank used to hold the electrolyte is constructed of plexiglass to twice the physical size of the specimen being modelled. This size increase helps to reduce the errors due to probe lead positioning. The current input leads, either in the form of a line or point source, are positioned either at the surface or to any depth in the electrolyte.

Figure 3 - Photo of Electrolyte Bath Analogue

Figure 4 shows the voltage ratio V/V_o, (i.e., cracked/uncracked voltage) for an a/w ratio of 0.5 in a centre-cracked specimen subjected to a constant, uniform current field. It can be noted that the sensitivity to crack growth is significantly reduced if the voltage probes are placed greater than approximately 5 mm from the crack plane. However, if an attempt is made to increase probe sensitivity by using positions close to the crack, errors will occur unless the crack and the distance to the probe are clearly defined. Probe position should be a compromise based on the current available, the level of the output signal, the resolution required and the accuracy of the probe positioning.

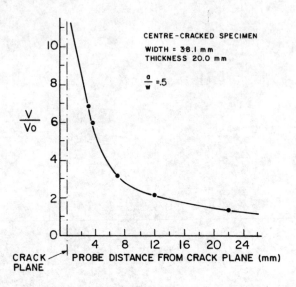

Figure 4 - Normalized Potential Drop versus Probe Distance from Crack Plane for Centre-Cracked Specimen

5. ONE-DIMENSIONAL CRACK GROWTH MONITORING

To test the ability of the electrolytic analogy as a calibration technique, it was used to predict the crack size of a centre-cracked tension specimen subjected to fatigue. The specimen was cycled and heat marked at a total of eight crack lengths to provide experimental points for comparison. This comparison is shown in Figure 5. Note that the maximum error occurs at the larger crack sizes and corresponds to an error of less than .3 mm (< 1.0%). This error is believed to be caused by the crack growing slightly nonuniformly due to a bending component in the loading. Other one-dimensional crack growth specimens with single edge notches have also been calibrated with similar results.

6. TWO-DIMENSIONAL CRACK GROWTH MONITORING

The use of Potential Drop measurements with the analogue calibration technique is very convenient for one-dimensional crack growth. It is also very convenient for two-dimensional crack growth such as the growth of surface defects if the analogue is again used for calibration. As described earlier, the measurement of the growth of surface cracks is obtained by using multiple probes spaced along the specimen and across the crack.

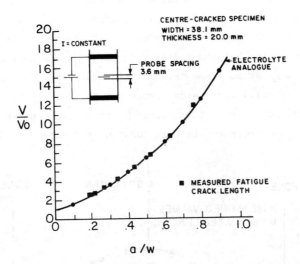

Figure 5 - Comparison of Measured and Predicted Crack Length using Analogue Technique for Centre-Cracked Specimen

In determining the defect profile at any instant one first attempts to define the surface crack length (2c). In many cases this can be determined optically. Otherwise a method that the authors have found to be reasonably accurate is to plot the probe voltage (normalized by the uncracked voltage) as a function of probe distance on the surface of the specimen (Figure 6). Referring to Figure 6, the crack length (2c) can be found from the points at which the potential becomes constant. From a line drawn through the points, a curve can be fitted for the normalized Potential Drop along the crack front. Using the value 2c and the calibration curves generated by the electrolyte analogue (Figure 8), the maximum crack depth at the centreline (a_{max}) can be evaluated. The remaining crack profile can then be estimated by using the relationship:

$$\frac{\left(\dfrac{V_i}{V_o} - \dfrac{V_E}{V_o}\right)}{\left(\dfrac{V_{max}}{V_o} - \dfrac{V_E}{V_o}\right)} \, a_{max} = a_i \, ,$$

where V_{max}/V_o = maximum normalized potential (at a_{max} for an elliptical crack); V_E/V_o = normalized potential across the crack plane in the uncracked region; V_i/V_o = normalized potential at the point where crack depth a_i is required. Using this technique, reasonable agreement was obtained between the measured and predicted crack profile of the defect shown in Figure 7. The maximum disagreement was approximately .5 mm.

It should be noted that the calibration curves shown here were generated assuming a relatively smooth, elliptically-shaped defect. While this assumption is valid for a wide number of applications, especially in fatigue, the applicability of this technique to predict the shape of irregularly shaped defects is still in question and currently under study.

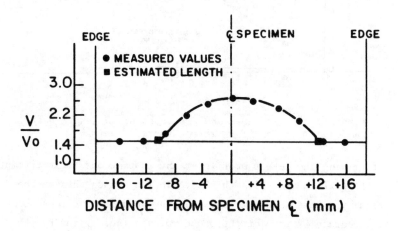

Figure 6 - Normalized Potential Drop versus Probe Distance on Specimen

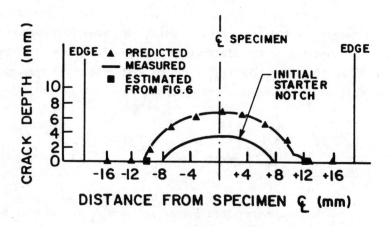

Figure 7 - Predicted and Measured Crack Depth versus Distance Across Specimen

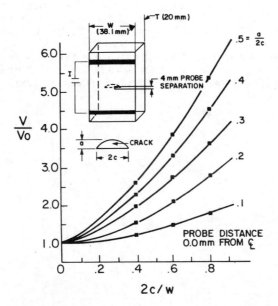

Figure 8 - Calibration Curves for an Elliptical Crack in a Finite Width Plate

7. CONCLUSIONS

The authors have found the Pulsed D.C. Potential Drop Method to be effective in measuring crack growth in both one and two dimensions. For two dimensional cracks an electrolytic bath analogue has considerably shortened the time required to calibrate the technique.

ACKNOWLEDGEMENTS

The authors would like to thank NOVA, An Alberta Corporation and the American Iron and Steel Institute for their support. The authors are grateful to M. Van Reenan of the Department of Mechanical Engineering, University of Waterloo for his design of the Potential Drop circuits.

REFERENCES

1. KNOTT, J.F., "The Use of Analogue and Mapping Techniques with Particular Reference to Detection of Short Cracks", *The Measurement of Crack Length and Shape During Fracture and Fatigue*, Chameleon Press Ltd., London, 1980.

2. PICK, R.J., GOVER, A.G. and COOTE, R.I., "Full Scale Testing of Large Diameter Pipelines", *Proceedings of the Welding Institute of Canada Conference on Pipeline and Energy Plant Piping*, Calgary, Alberta, November, 1980.

3. GLOVER, A.G. and COOTE, R.I., "Full-Scale Fracture Tests of Pipeline Girth Welds", presented at the *ASME 4th National Congress on Pressure Vessel and Piping Technology*, Portland, Oregon, 1983.

MODELLING PROBLEMS IN CRACK TIP MECHANICS   *Waterloo, Ontario, Canada*
CFC10, University of Waterloo   *August 24-26, 1983*

DUCTILE FAILURE OF HIGH TOUGHNESS LINE PIPE CONTAINING
CIRCUMFERENTIAL DEFECTS

D. ROMILLY, R.J. PICK and D.J. BURNS   R.I. COOTE
Department of Mechanical Engineering   Materials Engineering Services
University of Waterloo   NOVA, An Alberta Corporation
Waterloo, Ontario, Canada   Alberta, Canada

1.   INTRODUCTION

The acceptability of imperfections detected in pipeline girth welds is generally determined with respect to maximum allowed sizes based on workmanship standards for particular types of imperfections in pipeline standards such as CSA Z184, API 1104 and B.S. 4515.

In some cases, however, the acceptability may be determined on the basis of an Engineering Critical Assessment (ECA) using a fracture mechanics analysis. A series of full scale fracture tests in girth welds containing defects was designed and conducted in a joint program at the University of Waterloo and the Welding Institute of Canada under the sponsorship of NOVA, An Alberta Corporation, and subsequently by the American Gas Association. The majority of the tests were performed on 914 mm and 1067 mm diameter spiral welded pipe using a distributed bending load and have been reported in previous publications [1,2].

Previous analysis of these tests [2] showed that the methodology of British Standard PD 6493 was conservative by at least a factor of 2 on the strain at failure or defect depth, except for tests of longer defects in which ductile failure occurred by stable tearing through the remaining ligament.

Calculations have shown that in a number of these tests failure occurred near the maximum bending moment for an ultimate ductile failure of the pipe. To investigate this further, calculations have been made to relate failure to the reaching of the flow stress in part or all of the pipe. Calculations were also made for flat plate tensile tests undertaken by the authors, pipe and plate tests undertaken by Erdogan et al [3] and pipe tests undertaken by Wilkowski and Kanninen [4] and Wilkowski and Eiber [5].

2. NET SECTION FLOW

Net section flow can be defined as occurring when the average stress in the material at the plane of the defect reaches the flow stress. The flow load is then dependent on the definition of the flow stress and the dimensions of the specimen and defect. For materials such as line pipe steels with little strain hardening the flow stress can be reasonably assumed to be the average of the yield and ultimate stresses. For material with more significant strain hardening this definition of flow stress is undoubtedly more approximate in representing the average stress across the section.

In the following sections data from various sources is studied to indicate if failure of flat plates in tension and pipes in bending can be predicted from net section flow or similar behaviour.

3. FLAT PLATE TESTS

To examine the fracture behaviour of high toughness line pipe material (CSA Z245.1 Grade 483), a number of tensile and bend specimens were loaded at room temperature under displacement control until a maximum load was exceeded. The results indicate that the maximum load, P_L, corresponded to a condition of flow stress on the net section at the plane of the crack. These limit loads were calculated using the following relations for surface defects in tension:

(i) Elliptical crack, $P_L = \sigma_f (Wt - \frac{ac}{2})$,   (1)

(ii) Edge crack, $P_L = \sigma_f W(t-a)$,   (2)

(iii) Centre crack, $P_L = \sigma_f t(W-2c)$.   (3)

For surface defects in bending the relationship was:

(iv) $P_L = \sigma_f \, \gamma \, W(t-a)^2/4S$,   (4)

where P_L = limit load; σ_f = flow stress; W = total width; t = thickness; a = crack depth; 2c = crack length; S = span; γ = 1.543 (for sharp crack), or 1.261 (for blunt notch).

The results of the tests and analyses for flat plates are given in Figure 1. All of the results are either conservative or are within one percent of the predicted limit load. Further confirmation of this analysis is found by considering data recently published by Erdogan [3] for

elliptical surface cracks and edge cracks in API X70 pipe material,
(Figure 1), and in the results of Hopkins [6] for edge cracks in welded
tensile specimens. Note that all the results are close to the predicted
values leading to the observation that crack shape has little effect on
the maximum load.

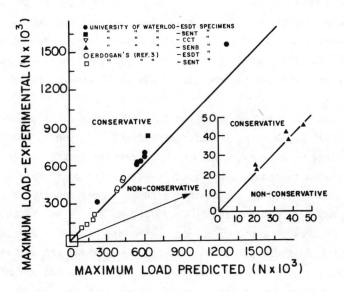

*Figure 1 - Experimental versus Predicted Maximum Loads for Small Specimens
of X70 Pipe Material*

4. PIPE TESTS

Bending tests of pipes containing defects have been undertaken by
Wilkowski and Kanninen et al [4], Wilkowski and Eiber [5], Erdogan [3]
and Glover and Coote [1,2].

In the tests by Kanninen et al [4], twenty specimens of 50.2, 101.4
and 406.4 mm diameter type 304 stainless steel pipe were tested in four
point bending with long circumferential defects. Both through wall and
internal circumferential cracks were studied. Maximum loads were predicted
using a net section collapse analysis based on the formation of a plastic
hinge in the plane of the defect. The correlation between the measured
and predicted maximum moments were good if corrections for the shift of
the neutral axis and ovalling were used and the flow stress was taken as
1.15 of the average of the yield and ultimate stresses.

Early initiation of stable crack growth was noted in the tests of 406.4 mm diameter, 26.2 mm thick pipe with long internal surface cracks. Kanninen et al [4] suggested that the prediction of maximum load was unconservative unless the defect size used included the crack growth. The authors believe that the predicted collapse moment was calculated using incorrect pipe dimensions and in fact the analysis is conservative. Thus one can conclude that the results of this study support the use of a net section flow analysis to predict the maximum moment. Questions remain, however, about the choice of flow stress for high strain hardening materia such as stainless steel and the role of stable crack growth, if any, in net section flow.

Wilkowski and Eiber [5] measured the failure stress for 100 and 150 mm diameter AISI 1020 pipes with circumferential machined grooves loaded in four point bending. The results were used to develop a stress magnification factor related to crack size. Failure was predicted if this magnification factor, multiplied by the nominal bending stress equalled the flow stress. This analysis was applied to three 762 mm x 15.9 mm API X60 pipes with machined grooves in girth welds and reasonable agreement was found with experiment. In these pipes the weld metal had yield and ultimate strengths of 573 and 658 MPa respectively compared to 410 and 557 MPa for the base metal, and agreement was achieved by assuming the flow stress was equal to the average of the base and weld metal yield strength plus 68.9 M This value, 560.4 MPa, is slightly higher than the average flow stress (549 MPa) of the weld and parent material where flow stress is defined as the average of the yield and ultimate strength.

Erdogan [3] tested six 508 mm diameter pipes containing circumferentia defects as well as the plate specimens described earlier. In addition Glover and Coote [2] report a series of fourteen tests on girth welds containing defects in 914 mm diameter pipe of various wall thicknesses. The tests of Glover and Coote indicated that brittle failure was unlikely except at temperatures below -45°C and as mentioned earlier, showed that ductile failure occurred by stable crack growth through the remaining ligament.

5. NET SECTION FLOW IN LARGE DIAMETER PIPE TESTS

The failure moments of the tests by Wilkowski and Eiber [5], Erdogan [and Glover and Coote [2] have been compared to three flow stress controlle criteria:

(a) the formation of a plastic hinge;

(b) the limit of elastic restraint in the region of the defect; and

(c) the stress magnification technique proposed by Wilkowski and Eiber [5].

5.1 Formation of a Plastic Hinge

The maximum moment, calculated as the moment at which the full section of the pipe in the plane of the crack reaches the flow stress, is compared to the experimentally measured collapse moment in Figure 2 for all the pipe data reported [2,3,4,5]. The flow stress used is the average of the yield and ultimate stress of the pipe or weld material at the plane of the crack. The flow stress for stainless steel was increased by a factor of 1.15. Points with arrows indicate termination of the test for reasons such as the local buckling of the pipe. It can be seen that for the large variety of defects and pipe sizes considered, the results are mainly unconservative. The most unconservative results being those tests where the nature of the flow stress was difficult to define accurately due to a mismatch in the weld and pipe material [5]. If such an approach were to be used to predict ultimate pipe moments a factor of safety or a reduced value of flow stress could be used to produce a conservative result.

5.2 Elastic Restraint Limit

The maximum bending moment in tests with surface defects is experimentally observed to correspond approximately to the point when the surface defects grows fully through the wall. Unless the original surface defect is short, the resulting through wall defect will exceed the critical length and unstable fracture will occur.

It is postulated that the formation of a through wall defect from a surface defect will occur when the ligament strain is not limited by the elastic restraint in the material at the ends of the defect, providing that premature brittle failure does not occur. Precise determination of the instant when the loss of elastic restraint occurs at the ends of the defect has not been attempted. However, the condition has been estimated to occur when the nominal stress in the full pipe wall at the ends of the defect reaches the flow stress, with the stress in the remaining pipe wall decreasing to zero at the neutral axis of the uncracked pipe. The stress in the original uncracked ligament is also taken to be the flow stress in calculating the estimated maximum moment. It can be seen in Figure 3 that the maximum moment is a conservative prediction of the failure moment except for three tests reported by Glover and Coote [2]. These three tests were specially prepared cracked welds and showed early initiation of stable crack growth. Glover and Coote suggest that weld residual stresses resulting from the method used to produce the crack may be different from normal welds and therefore lead to early crack initiation.

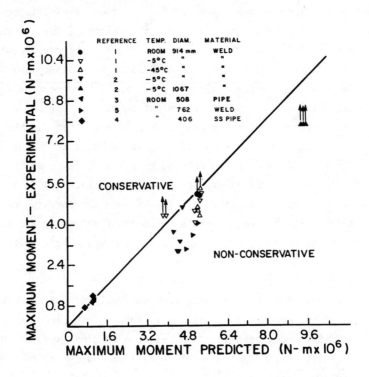

Figure 2 - Experimental versus Predicted Maximum Moments Using Plastic Hinge Analysis

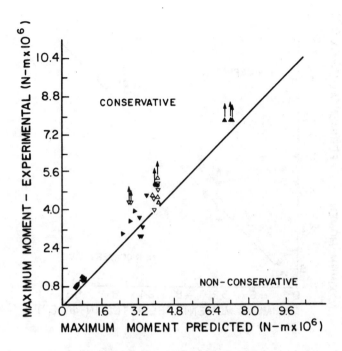

Figure 3 - Experimental versus Predicted Maximum Moments Using Elastic Restraint Limit Analysis

5.3 Wilkowski and Eiber Formula

The technique of Wilkowski and Eiber [5] making use of a stress magnification factor developed in small diameter pipe tests has also been used to predict the maximum moment. The comparison with experiment is shown in Figure 4. Again the results are all conservative except for the three tests of Glover and Coote.

6. DISCUSSION

The results of Wilkowski and Eiber are very similar to those predicted by the termination of elastic restraint suggesting that the empirical equation developed by Wilkowski and Eiber is in some way describing this condition.

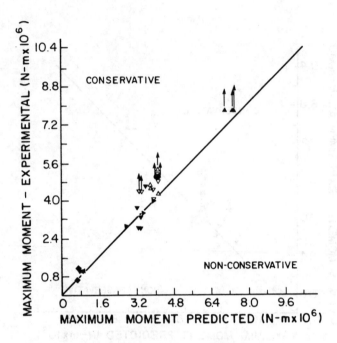

Figure 4 - Experimental versus Predicted Maximum Moments Using Battelle Empirical Analysis

For through thickness defects the equivalent longitudinal elastic stress at failure (σ_L) is given by Wilkowski and Eiber [5] as:

$$\sigma_L = \sigma_f/M_c ,$$

where

$$M_c = \left[1 + .26\left(\frac{c}{\pi R}\right) + 47\left(\frac{c}{\pi R}\right)^2 - 59\left(\frac{c}{\pi R}\right)^3\right]^{1/2} ,$$

and σ_f = flow stress; c = defect length; and R = pipe radius.

For surface defects

$$\sigma = \sigma_f/M_s ,$$

where

$$M_s = \left(1 - \frac{d/t}{M_c}\right)(1-d/t)^{-1} .$$

Figure 5 shows a plot for M_s for various lengths of surface cracks. It can be seen that as d/t approaches 1.0, M_s goes to infinite rather than equalling the value of M_c as should be expected. This indicates that there is an inconsistency in the definition of M_s.

A similar value of M_s can be computed for the situation of the ligament reaching the flow stress (termed the elastic restraint limit). It can be seen on Figure 5 that this agrees well with Wilkowski and Eiber for d/t less than 0.5 but disagrees for values greater than 0.5. The values of M_c also disagree for through thickness cracks.

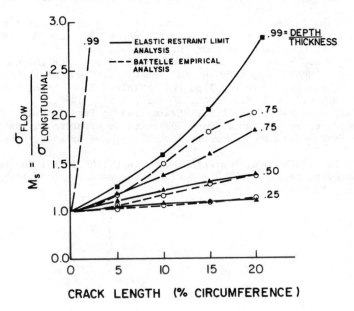

Figure 5 - Comparison of Derived Stress Magnification Factors for Two Analysis Techniques

7. CONCLUSION

The results and analysis presented indicate that the maximum bending moment for pipes containing circumferential defects can be accurately estimated from the stress distribution corresponding to the elastic restraint limit at the edges of the defect.

ACKNOWLEDGEMENTS

This study was supported by NOVA, An Alberta Corporation. The authors would like to thank the Welding Institute of Canada for their cooperation.

REFERENCES

1. PICK, R.J., GLOVER, A.G. and COOTE, R.I., "Full Scale Testing of Large Diameter Pipelines", *Proceedings of the Welding Institute of Canada Conference on Pipeline and Energy Plant Piping*, Calgary, Alberta, November, 1980.
2. GLOVER, A.G. and COOTE, R.I., "Full-Scale Fracture Tests of Pipeline Girth Welds", *ASME 4th National Congress on Pressure Vessel and Piping Technology*, Portland, 1983.
3. ERDOGAN, F., "Theoretical and Experimental Study of Fracture in Pipelines Containing Circumferential Flows", *Report No. DOT-RSPA-DMA-50/83/3*, U.S. Department of Transport, August, 1982.
4. KANNINEN, M.F., WILKOWSKI, G.M., ABOW-AAYEH, I., MARSCHALL, C., BROEK, D., SAMPATH, S., RHEE, H. and AHMAD, J., "Instability Prediction of Circumferentially Cracked Type 304 Stainless Steel Pipes under Dynamic Loading", *EPRI Report No. NP 2347*, April, 1982.
5. WILKOWSKI, G.M. and EIBER, R.J., "Evaluation of Tensile Failure of Girth Weld Repair Grooves in Pipe Subjected to Offshore Laying Stresses *ASME Conference on Energy-Sources Technology*, New Orleans, February, 1980.
6. HOPKINS, P., JONES, D.G. and FEARNEHOUGH, G.D., "Defect Tolerance in Pipeline Girth Welds", *ASME 4th National Congress on Pressure Vessel and Piping Technology*, Portland, 1983.

ON THE DEVELOPMENT OF A MIXED MODE TENSILE-SHEAR INTERLAMINAR FRACTURE
CRITERION FOR GRAPHITE-EPOXY COMPOSITES

A.J. RUSSELL and K.N. STREET
Composites Group
Defence Research Establishment Pacific
Victoria, British Columbia, Canada

1. INTRODUCTION

Along with the recent increase in the use of advanced composite materials
in primary aircraft structures, has come the need for a quantitative approach to inspection and damage tolerance of these materials. Not only
must allowable flaw limits be determined, but damage growth rates must also
be known in order to provide a rational basis for establishing inspection
intervals. There have been numerous efforts [1-9] to make use of linear
elastic fracture mechanics to describe the behaviour of flawed composite
materials. However, when fracture of the fibres is predominant, it is
found [1-4] that flaw propagation involves the growth of a damage affected
zone rather than the self-similar extension of a single crack as is normally the case in fracture mechanics. On the other hand where fibre
breakage is restricted such as in the longitudinal splitting of unidirectional composite materials or in the interlaminar fracture of laminated
composites, flaw growth is more likely to remain on a single fracture
plane [5-9].

The equilibrium and stability of delaminations are generally examined
in terms of the strain energy release rate G since G is well defined mathematically. and easily measured experimentally. Moreover, recent work [10-12]
has indicated that both finite element analysis as well as anisotropic
elasticity theory can be used to determine not only the total available
strain energy release rate but also the contribution from each of the
fracture modes, mode I, mode II and mode III. This latter information is
essential since experimental measurements have indicated [13,14] that the
critical strain energy release rate for fracture, or fracture energy G_c,
is strongly dependent on the fracture modes present. Not only must a
fracture criterion for delamination take into account the fracture modes
present, but must also take into account environmental effects such as

temperature and absorbed moisture content which are also known to influence the fracture energy [15,16].

2. MIXED MODE FRACTURE CRITERION

The strength of graphite-epoxy laminates is determined to a large extent by the size and distribution of the defects present. It is therefore a reasonable hypothesis that this response to microscopic defects together with the elastic properties might provide sufficient information to characterize the behaviour of macroscopic flaws. Such an approach was adopted by Wu [17] to describe the mode I/mode II splitting of unidirectional Scotchply 1002 fibreglass. For the case of delamination in graphite/epoxy, a fracture criterion mathematically the same as that of Wu has been employed although the nature of the stress state and evaluation of the strength parameters are different.

For an orthotropic material with the crack oriented in the principal direction the crack tip stress distribution can be expressed as [17,18]:

$$\sigma_i = \frac{k_1}{\sqrt{r}} \, g_i(a_{ij}, \theta) + \frac{k_2}{\sqrt{r}} \, h_i(a_{ij}, \theta) \, , \tag{1}$$

where k_1 and k_2 are the mode I and mode II stress intensity factors and the a_{ij} are the material elastic constants. The functions g_i and h_i can be obtained via the Airy stress function in a manner similar to the isotropic case [19]. For multiaxial loading the strength of composite materials is well described by the polynomial failure tensor [20,21] which has the form:

$$\Sigma \, F_i \sigma_i + \Sigma \, F_{ij} \sigma_i \sigma_j + \Sigma \, F_{ijk} \sigma_i \sigma_j \sigma_k + \cdots \leq 1 \, . \tag{2}$$

In the Wu model of mixed mode fracture, the crack tip stresses are used in conjunction with the failure tensor to determine a characteristic volume for fracture, r_c, which is defined as the greatest value of r in equation (1) such that the equality condition in equation (2) is met. The angle, θ_c, at which this occurs is termed the critical angle for fracture. The essence of the Wu fracture criterion is that the characteristic volume for failure, r_c, is constant for all combined loading conditions.

3. EXPERIMENTAL PROCEDURES

The strain energy release rate is directly related to the specimen compliance, C, via the equation:

$$G = \frac{P^2}{2} \frac{d}{da} \left(\frac{C}{b} \right),$$ (3)

where, P is the applied force; b, the specimen width and a, the crack length. For constant width and thickness specimens, the compliance can be expressed as a function of a,b,h and a_{ij} where 2h is the specimen thickness, i.e., $C = C(a,b,h,a_{ij})$. By differentiating w.r.t. 'a' and substituting C for the a_{ij} and h it is possible to express G as a function of a,b and C, i.e., $G = G(a,b,C)$, all of which can be measured experimentally.

The four mixed mode specimens used in the present work are shown in Figure 1 together with the calculated expressions for G. The pure mode I and mode II specimens have been described previously [22]. The compliance of the 43 percent mode II specimen is evaluated in the same way as the pure mode II specimen and the ratio is obtained from the compliance due to crack opening. The 83 percent mode II specimen has been described by Wilkins [23] except that he used a value of 76.5 percent for the mode II contribution based on a finite element analysis. A similar analysis by the authors using 20 noded isoparametric brick elements and 15 noded isoparametric crack tip elements resulted in the 83 percent mode II used in the present work.

Specimens, except for the 83 percent mode II specimen, were fabricated from 24 ply laminates, 3 mm thick; the 83 percent mode II specimens were only 6 ply thick. Teflon-coated glass fabric was positioned in the lay-up to act as the delamination starter and a crosshead speed of 2.5 mm/min. was used for all tests. The compliance measurements were made with the use of a clip gauge in the case of the mode I specimen, an LVDT in the case of the 3-point loaded specimens and strain gauges in the case of the 83 percent mode II specimen.

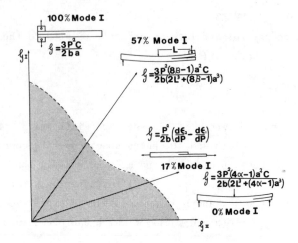

*Figure 1 - Schematic Showing Specimen Geometries, Percent of Mode I
Fracture and Expressions for Strain Energy Release Rates*

4.   RESULTS AND DISCUSSIONS

Figure 2 shows the experimental results obtained for unidirectional
AS-1/3501-6 graphite/epoxy at 20°C. Also shown are the predicted curves
assuming either plane stress or plane strain conditions. The elastic
properties and strength parameters used in the calculations are shown in
Table 1. These values, a combination of measured values and those reported
in the literature [24,25] are thought to be the best available for AS-1/
3501-6. The r_c value of 5 x 10^{-4} m, chosen to give a good fit to the
pure mode II data, also fits the mode I data well for plane strain but
overestimates G_c for plane stress under mode I dominated conditions. Moreover neither of the predicted curves agrees well with the data for the
83 percent mode II specimen.

As can be seen from Table 1, only first and second order terms have
been included in the failure tensor since for most applications these are
generally found to be sufficient. However, since σ_3 and σ_5 are the dominant
stresses in the present problem, the effect of including their lowest
order interaction term, F_{355}, was investigated. It was found, Figure 3,
that an F_{355} value of -0.15 x 10^{-6} (MPa)$^{-3}$ gave a good fit to all the data
for plane strain. No experimentally determined value of F_{355} has yet been
obtained for AS-1/3501-6, but the similar parameter F_{266} has been reported
[26] for another graphite/epoxy system as -0.69 x 10^{-6} (MPa)$^{-3}$.

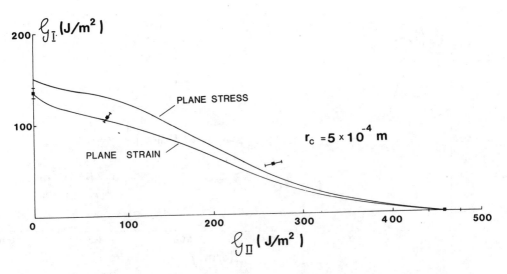

Figure 2 - Mixed Mode Fracture Energies for Unidirectional AS-1/3501-6 at 20°C, Dry and Predicted Curves for Plane Stress and Plane Strain. Error bars indicate standard error in the mean value.

Table 1 - Elastic Constants and Strength Parameters for AS-1/3501-6 Graphite/Epoxy at 20°C, Dry

Property	Value
F_1	-0.212×10^{-3} $(MPa)^{-1}$
F_2, F_3	16.8×10^{-3} $(MPa)^{-1}$
F_{11}	0.512×10^{-6} $(MPa)^{-2}$
F_{22}, F_{33}	81.4×10^{-6} $(MPa)^{-2}$
F_{55}	52×10^{-6} $(MPa)^{-2}$
F_{12}	3.2×10^{-6} $(MPa)^{-2}$
a_{11}	8.0×10^{-12} $(Pa)^{-1}$
a_{22}, a_{33}	100×10^{-12} $(Pa)^{-1}$
a_{55}	240×10^{-12} $(Pa)^{-1}$
a_{12}, a_{13}	-2.2×10^{-12} $(Pa)^{-1}$

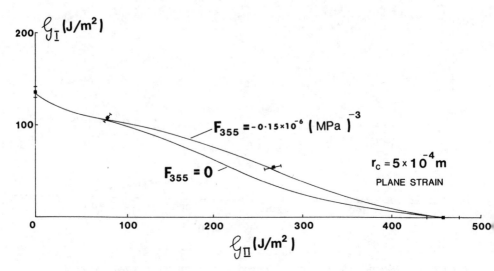

*Figure 3 - Mixed Mode Fracture Energies for Unidirectional AS-1/3501-6
at 20°C, Dry and the Effect of F_{355} Value on the Predicted
Curve for Plane Strain*

Figure 4 shows the change in fracture energy with stable crack extension for each type of specimen. While the mode II dominated fracture
energies remain constant, the other two specimens show a progressive increase in G_c. This increase with crack extension has been attributed to
fibres bridging or skipping across the crack. Figure 5 shows the effect
of fracture mode on the critical angle for fracture. It seems likely
that the large critical angle for the mode I dominated fracture results
in cracking out of the interlaminar plane whereas the near 0° angle for
mode II dominated fracture does not. On the premise that a bridged fibre
eventually results in a broken fibre the areal densities of fibre ends
on the fracture surfaces, n, were measured. For pure mode I, n = 35/mm^2
while for pure mode II, n = 9/mm^2 confirming the greater tendency for
fibre bridging in mode I. The contribution of bridged fibres to the total
fracture energy has been discussed in an earlier report and is found to be
strongly dependent on temperature, moisture content and specimen thickness [27].

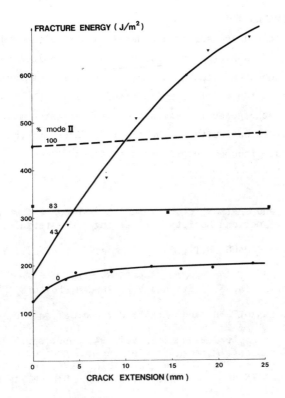

Figure 4 - *Effect of Crack Extension on the Mixed Mode Fracture Energies of AS-1/3501-6 at 20°C, Dry*

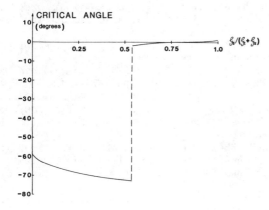

Figure 5 - *Effect of Fracture Mode on the Critical Angle for Fracture* θ_c

5. CONCLUDING REMARKS

The model presented appears to adequately describe the mode I/mode II interlaminar fracture behaviour of <u>unidirectional</u> graphite/epoxy laminates. However, several questions remain including the effect of environment on r_c, the extension of the model to include mode III and the applicability of the approach to the general delamination problem which must address the issues of elastic discontinuity at the ply interfaces and the effects of thermal and moisture induced interlaminar stresses.

REFERENCES

1. SLEPTZ, J.M. and CARLSON, L., *Fracture Mechanics of Composites*, ASTM STP 593, American Society for Testing and Materials, 1975, pp. 143-162.
2. MORRIS, D.H. and HAHN, H.T., *J. of Composite Materials*, Vol. 11, 1977, pp. 124-138.
3. SIH, G.C., *Proc. of the First USA-USSR Symposium on Fracture of Composite Materials*, G.C. Sih and V.P. Tamuzs, Editors, Sijthoff and Noordhoff, 1979.
4. BATHIAS, C., ESNAULT, R. and PELLAS, J., *Composites*, Vol. 12, 1981, pp. 195-200.
5. WU, E.M., *J. of Applied Mechanics*, Vol. 34, 1967, pp. 967-974.
6. WANG, S.S., *Composite Materials: Testing and Design*, ASTM STP 674, American Society for Testing and Materials, 1979, pp. 642-663.
7. BASCOM, W.D., BITNER, J.L., MOULTON, R.J. and SIEBERT, A.R., *Composites*, Vol. 11, 1980, p. 9.
8. KONISH, H.J., SWEDLOW, J.L. and CRUSE, T.A., *J. of Composite Materials* Vol. 6, 1972, pp. 114-124.
9. SCOTT, J.M. and PHILLIPS, D.C., *J. of Material Science*, Vol. 10, 1975, pp. 551-562.
10. JONES, R. and CALLINAN, R.J., *Progress in Science and Engineering of Composites*, T. Hayashi, Editor, The Japan Society for Composite Materials, Vol. 1, 1982, pp. 447-453.
11. RAMKUMAR, R.L., KULKARNI, S.V. and PIPES, R.B., *Report No. 76228-30*, Naval Air Development Centre, 1976.
12. WANG, S.S., *Proc. of the 21st AIAA/ASME Structures and Materials Conference*, Atlanta, Georgia, 1981, pp. 473-484.
13. McKINNEY, J.M., *J. of Composite Materials*, Vol. 6, 1972, p. 164.
14. SIDEY, F.G. and BRADSHAW, F.J., *Proc. of the Int. Conf. on Carbon Fibres, Their Composites and Applications*, The Plastics Institute, London, 1971, Paper 25.
15. MILLER, A.G., HERTZBERG, P.E. and RANTALA, V.W., *Proc. of the 12th National SAMPE Technical Conference*, 1980, pp. 279-293.
16. THOMSON, K.W. and BROUTMAN, L.J., *Polymer Composites*, Vol. 3, 1982, p. 113.
17. WU, E.M., *Composite Materials*, L.J. Broutman, Editor, Vol. 5, Academic Press, 1973, pp. 191-247.
18. SIH, G.C., PARIS, P.C. and IRWIN, G.R., *Int. J. of Fracture Mechanics*, Vol. 1, 1975, pp. 189-203.

19. IRWIN, G.R., *J. of Applied Mechanics, Trans. of ASME*, Vol. 24, 1957, p. 361.

20. WU, E.M., *Composite Materials*, G.P. Sendeckyj, Editor, Vol. 2, Academic Press, 1973, pp. 353-431.

21. TSAI, S.W. and WU, E.M., *J. of Composite Materials*, Vol. 5, 1971, p. 58.

22. RUSSELL, A.J. and STREET, K.N., *Progress in Science and Engineering of Composites*, T. Hayashi, Editor, The Japan Society for Composite Materials, Vol. 1, 1982, pp. 447-453.

23. WILKINS, D.J., *Report No. NAV-GD-0037*, Naval Air Systems Command, September, 1981.

24. GRIMES, G.C. and ADAMS, D.F., *Report No. NOR79-17*, Northrop Technical Report, October, 1979.

25. *Hercules Magnamite Graphite Fibres*, Hercules Incorporated, Magna, Utah, 1978.

26. TENNYSON, R.C., MacDONALD, D. and NANYARO, A.P., *J. of Composite Materials*, Vol. 12, 1978, pp. 63-75.

27. RUSSELL, A.J., *Materials Report No. 82-Q*, Defence Research Establishment Pacific, December, 1982.

MODELLING PROBLEMS IN CRACK TIP MECHANICS   *Waterloo, Ontario, Canada*
CFC10, University of Waterloo   *August 24-26, 1983*

THE INFLUENCE OF NONSINGULAR STRESSES ON EXPERIMENTAL MEASUREMENTS
OF THE STRESS INTENSITY FACTOR

R.J. SANFORD
Department of Mechanical Engineering
University of Maryland
College Park, Maryland, U.S.A.

1. INTRODUCTION

It is well established that, within the framework of linear elastic fracture mechanics, the state of stress in the immediate vicinity of a crack tip under opening-mode loading can be represented by the modified near-field equations:

$$\sigma_x = \frac{K}{\sqrt{2\pi r}} \cos \frac{\theta}{2} \left(1 - \sin \frac{\theta}{2} \sin \frac{3\theta}{2}\right) + \sigma_{ox} \,,$$

$$\sigma_y = \frac{K}{\sqrt{2\pi r}} \cos \frac{\theta}{2} \left(1 + \sin \frac{\theta}{2} \sin \frac{3\theta}{2}\right) \,, \tag{1}$$

$$\tau_{xy} = \frac{K}{\sqrt{2\pi r}} \cos \frac{\theta}{2} \sin \frac{\theta}{2} \cos \frac{3\theta}{2} \,.$$

The term, σ_{ox}, represents a constant stress parallel to the crack line and was first introduced by Irwin [1] to account for the orientation of the isochromatic fringes around the crack-tip in the photoelastic experiments of Wells and Post [2]. The need to include this term in other aspects of near-field analysis is now widely accepted [3,4,5]. Unfortunately, the size of the region around the crack tip for which equations (1) adequately describe the state of stress is very small. In specimens of finite dimensions this region is of the order of one percent of the remaining uncracked ligament [6]. For practical reasons, experimental measurements of the stress field to determine the stress intensity factor must be made beyond the limits of this "singularity-dominated-zone" and, accordingly, the influence of nonsingular stresses on the overall stress state must be considered.

It has been shown [7] that a generalization of the Westergaard equations [8] is necessary for a complete representation of the stress state associated with two-dimensional cracks under static opening-mode loading.

The resulting expressions for the Cartesian stress components are:

$$\sigma_x = \sum_{n=o}^{n=N} A_n r^{n-1/2} [\cos(n-1/2)\theta - (n-1/2)\sin\theta\sin(n-3/2)\theta]$$
$$+ \sum_{m=o}^{m=M} B_m r^m [2\cos(m\theta) - m\sin\theta\sin(m-1)\theta] ,$$

$$\sigma_y = \sum_{n=o}^{n=N} A_n r^{n-1/2}[\cos(n-1/2)\theta + (n-1/2)\sin\theta\sin(n-3/2)\theta]$$
$$+ \sum_{m=o}^{m=M} mB_m r^m \sin\theta\sin(m-1)\theta , \qquad\qquad (2)$$

$$\tau_{xy} = \sum_{n=o}^{n=N} - (n-1/2)A_n r^{n-1/2}\sin\theta\cos(n-3/2)\theta$$
$$- \sum_{m=o}^{m=M} B_m r^m [m\sin\theta\cos(m-1)\theta + \sin(m\theta)] .$$

Techniques to determine the values of the unknown, real coefficients, A_n and B_m, in the above equations have been developed by the author and his co-workers in other papers [9,10,11,12] and will not be discussed here; rather the aim of this paper is to discuss the influence of these non-singular stress terms on the measurement of the opening mode stress intensity factor by a variety of experimental methods.

2.   PHOTOELASTICITY

The photoelastic method produces fringes which are contours of constant maximum shear stress, τ_m, where:

$$\tau_m^2 = \frac{1}{4} (\sigma_x - \sigma_y)^2 + \tau_{xy}^2 . \qquad\qquad (3)$$

Substituting equations (2) into equation (3) results in an equation which is highly nonlinear in the coefficients, which would appear to make the determination of K exceedingly difficult if the higher order effects cannot be neglected. Fortunately, the effect of each of the coefficients on the shape of the fringe pattern is distinct and each coefficient imparts a characteristic feature to the overall pattern. Representative photo-elastic fringe patterns from several fracture specimen geometries are shown in Figure 1. The coefficients of the nonsingular terms in each case are

quite different [6,13] as are the fringe patterns. An algorithm which
uses a large number of data points to characterize the fringe field allows
the coefficients to be extracted with relative ease. Experience indicates
that, except for unusual circumstances, only six to eight coefficients are
needed to determine the stress intensity factor to within one percent [13].
This is in marked contrast to numerical methods, such as the boundary
collocation method, in which 150 or more terms are required to achieve
this level of accuracy.

Modified Compact Tension (MCT)   *Three Point Bend (TPB)*

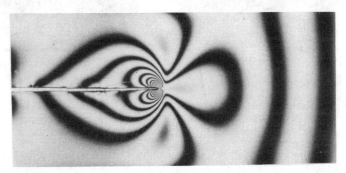

Rectangular Double Cantilever Beam (RDCB)

Figure 1 - Isochromatic Patterns in Various Fracture Specimens

3.   INTERFEROMETRY

Classical methods of optical interferometry or holographic interferometry
both result in fringe patterns which are contours of the constant sum of
the normal stresses (isopachics). Unlike the governing relation for the
isochromatic fringe pattern, the defining equation for the isopachic pattern
is linear in the nonsingular coefficients. As a consequence, the

influence of each term on the shape of the pattern is small. For
example, Figure 2 compares the isopachic patterns for a modified compact
tension specimen at an a/W of 0.8, based on near-field and six-parameter
representations. Note that, although there is a difference in the aspect
ratio of the fringes, the basic character of the two patterns is the
same. This complicates efforts to separate the influence of the singular
and nonsingular terms in order to accurately determine the stress intensity factor from patterns such as these.

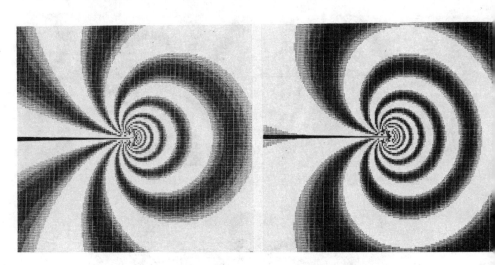

*Figure 2 - Isopachic Patterns at a/W = 0.8 in an MCT Specimen Based on
            Near-Field (left) and Six-Parameter (right) Stress Field
            Representations*

4. CAUSTICS

     When light passes through or is reflected from the surface of a speci-
men containing a crack, the localized anti-plane deformation around the
crack tip acts as a negative lens and refracts the light. The result is
a shadow-spot bounded by a bright, thin band of light, i.e., a caustic,
whose size is related to the stress intensity factor. Although funda-
mentally different in origin, the shape of the caustic is similar to that
of an isopachic fringe. Unlike an isopachic pattern, the caustic pattern
consists of a single "fringe" (or two fringes, for transmission through a
photoelastic material) from which data can be taken. This limited data

field severely restricts the experimentalists' ability to develop techniques for separating the influence of the singular and nonsingular stresses.
The usual procedure for analyzing a caustic pattern [14] is to measure the
size of the caustic in the direction normal to the crack, with K being
proportional to this measurement raised to the 2.5 power. Consequently,
even small changes in caustic size due to nonsingular effects give rise
to large changes in K. A systematic study of the influence of the nonsingular stresses on the size and shape of caustics has demonstrated that
the errors associated with neglecting nonsingular effects can be large
[15]. For example, Figure 3 compares the caustics based on near-field
and six-parameter representations for the same specimen geometry and
crack length as Figure 2. The correction factor given is a measure of the
error in the computed K-value and for the case shown is of the order of
9 percent.

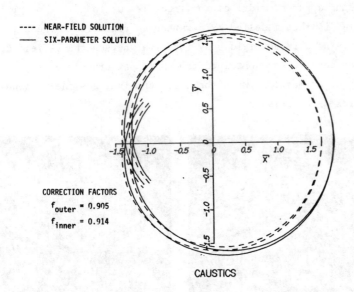

CAUSTICS

Figure 3 - Anisotropic Caustic Patterns Based upon Near-Field and Six-
Parameter Models at a/W = 0.8 in an MCT Specimen

5. MOIRE

The moire method provides contours of the component of displacement
along a chosen direction. The usual practice is to place the grid lines

perpendicular or parallel to the crack line so as to obtain the u or the
v displacement fields, respectively. Until recently, the moire method was
used primarily to investigate elasto-plastic behaviour because of its
limited sensitivity. With the development by Post [16] of the high-sensi-
tivity virtual reference grating method, measurements of elastic displacements can be made with relative ease.

As was the case for isopachics, the generalized equations for the displacements (obtained from equations (2)) are linear in the unknown coefficients and the approach for determining the coefficients is similar. There
are several differences however, which complicate the analysis of displacement data. First, the displacement field does not contain a dominant singularity term; and second, unless complex loading fixtures are used, the
recorded pattern frequently contains an additional rigid body rotation
which complicates the analysis. For example, Figure 4 shows u and v displacement fields for a modified compact tension specimen at an a/W or 0.6.
(Note that the apparent width of the crack is due to a lack of grating;
the crack was in fact modelled by a rather narrow slit.) The u-field
pattern contains a large rigid body rotation whereas the v-field does not.
However, the coefficients obtained from both patterns agree with those
obtained by other methods, when the algorithm includes the unknown rigid
body rotation as a variable.

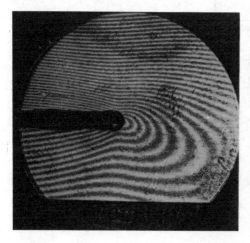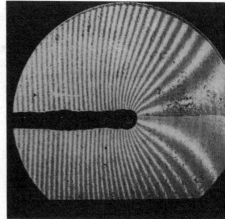

Figure 4 - Moire Displacement Patterns in the u-direction (left) and
v-direction (right) at a/W = 0.6 in an MCT Specimen

6. DISCUSSION

Full-field methods of experimental stress analysis can provide accurate determinations of the stress intensity factor. The choice of method depends upon the particular problem and the investigator's ability to account for all of the variables in the experiment. For practical reasons, additional parameters enter into the analysis and need to be evaluated in order to separate the influence of the singular term from other variables that affect the pattern. Experience gained from analyzing patterns obtained from the methods described indicates that there is a significant improvement in the accuracy of K measurement when additional terms are included in the series expansion of equations (2). It is suggested that, at the very least, terms through $r^{1/2}$ should be retained in any algorithm used to determine K from experimental mechanics data. Cracks approaching boundaries present particular problems and additional terms may be required in these cases.

REFERENCES

1. IRWIN, G.R., "Discussion of: The Dynamic Stress Distribution Surrounding a Running Crack - A Photoelastic Analysis", *Proceedings SESA, XVI*, 1958, pp. 93-96.
2. WELLS, A.A. and POST, D., "The Dynamic Stress Distribution Surrounding a Running Crack - A Photoelastic Analysis", *Proceedings SESA, XVI*, 1958, pp. 69-92.
3. SIH, G.C., "On the Westergaard Method of Crack Analysis", *Int. J. of Fracture Mechanics*, Vol. 2, 1966, pp. 628-631.
4. EFTIS, J. and LIEBOWITZ, H., "On the Modified Westergaard Equations for Certain Plane Crack Problems", *Int. J. of Fracture Mechanics*, Vol. 8, 1972, pp. 383-392.
5. LARSSON, S.G. and CARLSSON, A.J., "Influence of Non-singular Stress Terms and Specimen Geometry on Small-Scale Yielding at Crack Tips in Elastic-Plastic Materials", *J. Mech. Phys. Solids*, Vol. 21, 1973, pp. 263-277.
6. CHONA, R., IRWIN, G.R. and SANFORD, R.J., "The Influence of Specimen Size and Shape on the Singularity-Dominated Zone", *ASTM STP 791*, 1983, pp. I-3 - I-23.
7. SANFORD, R.J., "A Critical Re-examination of the Westergaard Method for Solving Opening-Mode Crack Problems", *Mechanics Research Comm.*, Vol. 6, 1979, pp. 289-294.
8. WESTERGAARD, H.M., "Bearing Pressures and Cracks", *Trans. ASME*, Vol. 61, 1939, pp. A49-A53.
9. SANFORD, R.J. and DALLY, J.W., "A General Method for Determining Mixed-Mode Stress Intensity Factors from Isochromatic Fringe Patterns", *Engineering Fracture Mechanics*, Vol. 11, 1979, pp. 621-633.
10. SANFORD, R.J., "Application of the Least-Squares Method to Photoelastic Analysis", *Experimental Mechanics*, Vol. 20, 1980, pp. 192-197.

11. IRWIN, G.R., et al, "Photoelastic Studies of Damping, Crack Propaga-
    tion, and Crack Arrest in Polymers and 4340 Steel", *NUREG/CR-1455*,
    University of Maryland, 1980.
12. SANFORD, R.J., CHONA, R., FOURNEY, W.L. and IRWIN, G.R., "A Photo-
    elastic Study of the Influence of Non-Singular Stresses in Fracture
    Test Specimens", *NUREG/CR-2179, (ORNL/Sub-7778/2)*, University of
    Maryland, 1981.
13. WHITMAN, G.D. and BRYAN, R.H., "Heavy-Section Steel Technology Pro-
    gram Quarterly Progress Report for April-June, 1982", *NUREG/CR-2751/
    Volume 2, (ORNL/TM-8369/V2)*, 1982, pp. 47-52.
14. BEINERT, J. and KALTHOFF, J.F., "Experimental Determination of
    Dynamic Stress Intensity Factors by Shadow Patterns", *Mechanics of
    Fracture VII*, edited by G.C. Sih, Martinus Nijhoff, 1981, pp. 281-330
15. PHILLIPS, J.W. and SANFORD, R.J., "Effect of Higher-Order Stress
    Terms on Mode-I Caustics in Birefringent Materials", *ASTM STP 743*,
    1981, pp. 387-402.
16. POST, D., "Developments in Moire Interferometry", *Optical Engineering*,
    Vol. 21, No. 3, 1982, pp. 458-467.
17. BARKER, D.B., SANFORD, R.J. and CHONA, R., "Stress Intensity Factor
    Determination from Displacement Fields", *Proc. of the SESA 1983 Annual
    Spring Meeting*, Cleveland, Ohio, 1983, pp. 445-448.

MODELLING PROBLEMS IN CRACK TIP MECHANICS   *Waterloo, Ontario, Canada*
CFC10, University of Waterloo   *August 24-26, 1983*

MEASUREMENTS OF NEAR TIP FIELD NEAR THE RIGHT ANGLE INTERSECTION
OF STRAIGHT FRONT CRACKS

C.W. SMITH and J.S. EPSTEIN
Department of Engineering Science and Mechanics
Virginia Polytechnic Institute and State University
Blacksburg, Virginia, U.S.A.

1. INTRODUCTION

One of the most difficult and controversial areas of research in analytical fracture mechanics has been the evaluation of the distribution of
stress intensification as a crack approaches a free boundary [1,2,3].   In
fact, three dimensional analysis suggests the need for the definition of
a quantity other than the classical stress intensity factor (SIF) for such
cases at the free surface of a body.

For over a decade, the first author and his associates have worked
towards the development of a photoelastic frozen stress modelling technique
for estimating SIF distributions in crack body problems where the SIF
varied along the flaw border. First proposed for Mode I loading only [4],
it was extended to all three modes of near tip deformation [5] and later
was found to predict flaw shapes under certain conditions [6]. Recently,
it was combined with a moire interferometric technique developed by Post
[7] to estimate near tip displacements as well. Preliminary results [8]
have suggested the feasibility of the current study. The present paper
deals with the use of the frozen stress-moire approach to study the variation in SIF distributions along cracks in finite thickness beams approximating ASTM fracture toughness beam specimens in geometry. Special attention has been given to the region where the crack front intersects the
free surface of the beam. Results are compared with Benthem's [3] analysis.

2. METHODS OF ANALYSIS

The experimental methods of analysis have been described in some detail
elsewhere [9,10,11] so only an abbreviated review is included here. The
specimen configuration and loading is given in Figure 1 and closely follows
the ASTM E399 [12] specification for size and shape requirements but also
including a very shallow flaw. Basically, two separate methods are used to

obtain SIF's. First, three dimensional (3-D) frozen stress photoelasticit
is used to obtain a stress derived SIF [9]. Second, the stress frozen
specimens are annealed with high density moire grids on them [8] to obtain
displacement derived SIFs. These separately obtained SIFs are then
compared.

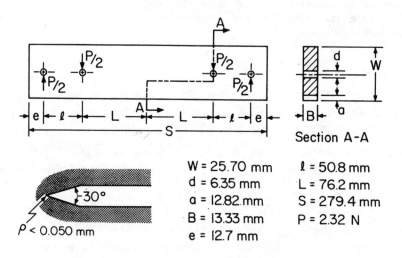

$$W = 25.70 \text{ mm} \qquad \ell = 50.8 \text{ mm}$$
$$d = 6.35 \text{ mm} \qquad L = 76.2 \text{ mm}$$
$$a = 12.82 \text{ mm} \qquad S = 279.4 \text{ mm}$$
$$B = 13.33 \text{ mm} \qquad P = 2.32 \text{ N}$$
$$e = 12.7 \text{ mm}$$

Figure 1 - Specimen Geometry (a/W = 0.5) from ASTM-E399

3.   RESULTS OF EXPERIMENTS

The procedures described in the preceding section were applied as indi
cated to the geometry of Figure 1 for three beams containing cracks of two
different relative depths. Pertinent dimensions and corresponding experimental SIF values are given in Table 1.

Table 1

a	w	B	a/W	W/B	$K_I/K_{Th}*$ Photoelastic	Moire
3.18	51.0	13.3	0.063	3.83***	1.06	1.04
12.70	25.4	13.3	0.500	1.91	1.01**	1.07**

```
*     K_Th is ASTM E399 2-D Solution
**    Average of Two Specimens
***   Too Thin for E399 Specifications
```

The results of the experiments for the deep cracks (a/W ≈ 0.50) are shown in Figure 2 and the curve is similar in shape to the curve for a/W = 0.063. These results show that both the photoelastic and moire methods yield SIF values which are close to but slightly above the two dimensional analytical solution in the central part of the beam. For an incompressible material ($\nu = 0.5$) along $\theta = \pi/2$, the two-dimensional near tip linear elastic fracture mechanics displacement field equations reveal that displacements associated with a state of plane strain are approximately 0.6 times the values associated with plane stress. This would seem to imply that, for a cracked model of sufficient thickness to produce a state of plane strain near mid-thickness, then the near tip displacements at the free surface should be about 60 percent of the mid-thickness values. However, Figure 2 suggests that this ratio is substantially less than 0.60. This result will be considered further in the sequel.

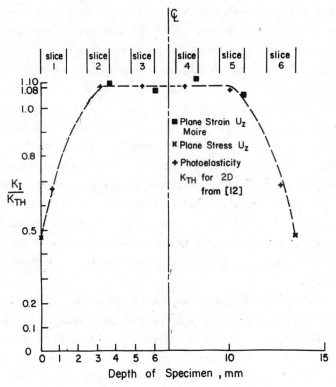

Figure 2 - Classical SIF Distributions from Photoelastic and Moire Data

4.   DISCUSSION

As noted above, the value of K_I at the free surface obtained from the
moire method using a plane stress assumption was much less than expected
and this was true for all three experiments. In seeking an explanation,
we turn now to a consideration of three dimensional analytical studies
which include the region where a crack intersects a free boundary.

Williams [13,14] obtained the stress singularities resulting from
various boundary conditions at angular corners of plate wedges under extension. His analysis led to an eigenvalue equation for the quantity λ_σ which
is the exponent of r in the expressions for the in-plane stresses. His
boundary conditions yielded a solution that led to the well known value
λ_σ = -1/2 for the lowest order of λ_σ for a crack. Subsequently, Sih [15]
pointed out that the value of λ_σ at the boundary was not known for three
dimensional cracked body problems and, as Benthem [16] noted, a state of
plane strain in a free-free wedge of angle 2π cannot be present near the
intersection of a crack front normal to the boundary of a half space unless
Poisson's Ratio vanishes. Additional studies, as noted earlier, [1,2,3]
and by Bazant [17] all confirm that λ_σ must be different at a free boundary
in three dimensional problems and it appears that $\lambda_\sigma \neq$ -1/2 (λ_u = 1-λ_σ,
the exponent of r in the in-plane displacements u_i) at the boundary surface.
Benthem's solution [16] for a quarter infinite crack in a half space
predicts a value of λ_u = 0.67 for ν = 0.50 and λ_u = 0.55 for ν = 0.30 at
the free surface. The mathematical correctness of these results were
subsequently confirmed [3] using an entirely different method of analysis.

In view of the above results, and the fact that the value of Poisson's
Ratio for stress freezing photoelastic materials approaches one half above
critical temperature, it was decided to devise a simple scheme for evaluating
the experimental data across the portion of the thickness through which λ
is believed to vary.

Agreement to within an experimental scatter of about ± 3 percent between
the photoelastic and plane strain moire results and a value about 8 percent
above the classical (λ_σ = -1/2) two dimensional solution in approximately
the central half of the plate is shown in Figure 2. We turn now to the
interpretation of the experimental data outside of the above noted central
region of the beam and how this data might be related to conventional
classical concepts regarding stress intensity factors. We begin by noting

that the moire value for K_1 plotted at the outer surface of the beam
(i.e., $K_I/K_{Th} \approx 0.46$) is incorrect since $\lambda_u = 1/2$ was used in the near tip
plane stress displacement field there. Moreover the photoelastic value
from the surface slice ($K_1/K_{Th} \approx 0.66$) is also incorrect, again because
$\lambda_\sigma = -1/2$ was used.

In order to obtain a better estimate in this region and to effect an
approximate connection with classical concepts, we have proceeded as described in the sequel. The moire displacement measurement at the surface
of the body is made on a free surface and, consequently, a state of plane
stress should exist there. Thus, it would seem that two dimensional (2-D)
concepts might be employed in describing the projection of the three
dimensional problem onto the 2-D free boundary.

Following [18] we write, for $\theta = \pi/2$,

$$u_z = C'(K_\lambda)_{AP} r^{\lambda_u} . \qquad (1)$$

However $(K_\lambda)_{AP}$ is defined by equation (1) and is not a classical stress
intensity factor since it is defined by r^{λ_u} rather than by $r^{1/2}$. Next we
relate $(K_\lambda)_{AP}$ to a pseudo classical $(K_p)_{AP}$ by

$$(K_p)_{AP} = (K_\lambda)_{AP} r^{(\lambda_u - 1/2)} . \qquad (2)$$

Equation (2) allows interpretation of the data as a classical apparent
stress intensity factor even though such a quantity does not exist in the
region near the free boundary.

In order to modify the boundary value of the SIF according to the
above scheme, we first determine λ_u as the slope of a log-log plot of u_z
versus r from (equation (1)). Next we compute values of $(K_\lambda)_{AP}$ from
equation (1). Finally, values of K_{AP} are computed from equation (2) and
plotted versus r^{λ_u}. By extrapolating the linear data to the origin, we
obtain what we may call a psuedo K_I or K_p at the free surface. In using
this approach, it is important to identify the proper zone within which
equation (1) is valid [9,11,19]. Figure 3 depicts this process graphically.

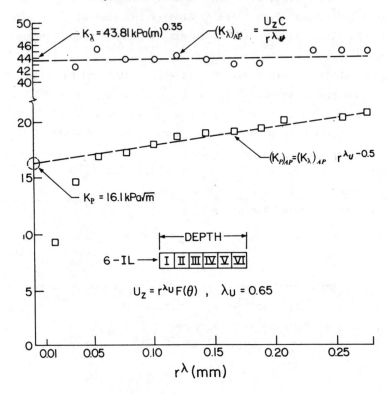

Figure 3 - Adjustment of Classical SIF Values near Boundary Surface in Accordance with 3-D Effects

In order to obtain a first estimate for the psuedo K_I or K_p distribution across the boundary layer thickness, we assumed linear upper and lower bounds represented by curves 1 and 3 in Figure 4. Then curve 2 was faired midway between the bounding curves 1 and 3 and tangent to the horizontal at its uppermost point. Although this operation is quite approximate, the error bound is of the order of ± 7 percent for the photoelastic point's λ_σ. This curve adjusts the surface slice photoelastic K_I from 0.66 to 0.91. Studies are currently underway in which this region is being further subdivided for more accurate moire readings.

From a log-log plot of equation (1), the value of λ_u for the region where the crack intersected the free surface was found to be 0.65 which

compares favourably with Benthem's value of 0.67. The adjustment of the
photoelastic surface slice value is of course less accurate.

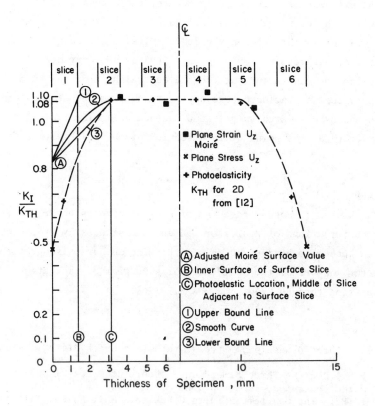

Figure 4 - Conversion of Optical Data Containing the 3-D Effects into K_p
*   Values near the Free Boundary*

5. SUMMARY

A series of integrated frozen stress-moire experiments were conducted
on stress freezing photoelastic models simulating ASTM E-399 Compact
Bending Specimens containing shallow and standard artificial cracks.
Using classical linear elastic fracture mechanics algorithms, estimates
of K_I were extracted from the optical data and compared with an analytical
two dimensional solution. Reasonable agreement between frozen stress results,
moire plane strain results and the two dimensional analytical solution was
found in the central half of the beam cross-section. Outside this region,

the classical K_I values fell well below the two dimensional analytical results. Accurate measurements on the outer surface of the beam yielded values of the exponent of r in the displacement field equations which agreed closely with the three dimensional analysis of Benthem. Computation of K_p values in this region yielded results significantly higher than classical planar results for $\nu = 0.5$. Studies are currently underway to better define the conditions outside the central half of the beam. The preliminary results included here appear to confirm the studies of Benthem, Sih and others which suggest the absence of a classical $\lambda_\sigma = -1/2$ stress singularity at the point of intersection of the crack with a free boundary.

ACKNOWLEDGEMENTS

The authors wish to acknowledge the pioneering efforts of Professor Folias, the contributions of Professor Sih and the studies by Professor Benthem which were most helpful to this work. The authors also gratefully acknowledge the support of the Solid Mechanics Program of the National Science Foundation through Grant No. MEA 811-3565.

REFERENCES

1. SIH, G.C., "Three Dimensional Stress State in a Crack Plate", *AFFDL TR-70-144*, 1970, pp. 175-191.
2. FOLIAS, E.S., "On the Three Dimensional Theory of Crack Plates", *J. of Applied Mechanics*, 1975, pp. 663-674.
3. BENTHEM, J.P., "The Quarter Infinite Crack in a Half Space: Alternative and Additional Solutions", *Int. J. of Solids and Structures*, Vol. 16, 1980, pp. 119-130.
4. SMITH, C.W., "Use of Three Dimensional Photoelasticity and Progress in Related Areas", *Experimental Techniques in Fracture Mechanics*, 2, *SESA Monograph*, edited by A.S. Kobayashi, Chapter 1, 1975, pp. 3-58.
5. SMITH, C.W., "Use of Photoelasticity in Fracture Mechanics", *Experimental Evaluation of Stress Concentration and Intensity Factors - Mechanics of Fracture*, edited by G.C. Sih, Chapter 7, Vol. 2, 1981, pp. 163-187.
6. SMITH, C.W. and PETERS, W.H., "Prediction of Flaw Shapes and Stress Intensity Distributions in 3D Problems by the Frozen Stress Method", *Preprints, Sixth Int. Conf. on Experimental Stress Analysis*, Munich, 1978, pp. 861-864.
7. NICOLETTO, G., POST, D. and SMITH, C.W., "Moire Interferometry for High Sensitivity Measurements in Fracture Mechanics", *Proc. of the Joint SESA-JSME Conf. on Experimental Mechanics*, May, 1982, pp. 266-270.
8. NICOLETTO, G., POST, D. and SMITH, C.W., "Experimental Stress Intensity Distributions in Three Dimensional Cracked Body Problems", *Proc. of Joint SESA-JSME Conf. on Experimental Mechanics*, May, 1982, pp. 196-200.

9. SMITH, C.W., "Use of Photoelasticity in Fracture Mechanics", *Experimental Evaluation of Stress Concentration and Intensity Factors (Mechanics of Fracture 7)*, edited by G.C. Sih, Chapter 2, Martinus-Nijhoff Publishers, 1981, pp. 163-188.

10. POST, D., "Optical Interference for Deformation Measurements - Classical, Holographic and Moire Interferometry", *Mechanics of Non-Destructive Testing*, edited by W.W. Stinchcomb, Plenum Publishing Co., 1980, pp. 1-53.

11. SMITH, C.W., "Use of Optical Methods in Stress Analysis of Three Dimensional Cracked Body Problems", *J. of Optical Engineering*, Vol. 21, No. 4, 1982, pp. 696-703.

12. "Standard E399 on Fracture Toughness Testing", *Annual Book of ASTM Standards, Part 10 Metals*, 1981, pp. 605-607.

13. WILLIAMS, M.L., "Stress Singularities Resulting from Various Boundary Conditions in Angular Corners of Plates in Extensions", *J. of Applied Mechanics*, Vol. 19, 1952, pp. 526-528.

14. WILLIAMS, M.L., "On the Stress Distribution at the Base of a Stationary Crack", *J. of Applied Mechanics*, Vol. 24, 1957, pp. 109-114.

15. SIH, G.C., "A Review of the Three-Dimensional Stress Problem for a Cracked Plate", *International J. of Fracture Mechanics*, Vol. 7, 1971, pp. 39-61.

16. BENTHEM, J.P., "State of Stress at the Vertex of a Quarter-Infinite Crack in a Half Space", *Int. J. Solids and Structures*, Vol. 13, 1977, pp. 479-492.

17. BAZANT, Z.P., "Three Dimensional Harmonic Functions Near Termination or Intersection of Gradient Singularity Lines; A General Numerical Method", *Int. J. of Engineering Science*, Vol. 12, 1974, pp. 221-243.

18. EFTIS, J., SUBRAMONIAN, N. and LIEBOWITZ, H., "Crack Border Stress and Displacement Equations Revisited", *J. of Engineering Fracture Mechanics*, Vol. 9, 1977, pp. 189-210.

19. LIU, H.W. and KE, J.S., "Moire Method", *Experimental Techniques in Fracture Mechanics 2*, edited by A.S. Kobayashi, Chapter 2 of SESA Monograph No. 2, 1975, pp. 111-165.

PLASTIC COLLAPSE STRESSES OF PLATES WITH SURFACE GROOVES

K.C. WANG, G. ROY and W.R. TYSON
Engineering and Metal Physics Section
Canada Centre for Mineral and Energy Technology
555 Booth Street
Ottawa, Ontario, Canada

1. INTRODUCTION

This paper forms part of an investigation into the prediction of failure stresses of damaged linepipe.

During pipelaying operations, pipe is sometimes damaged by heavy equipment. Contact can dent a pipe and score its surface. It is generally accepted that the most detrimental kind of surface damage is a dent containing a gouge.

The objective of the present work is to study the effect of a surface groove on the failure stress of linepipe. To complement parallel studies on gouged full-scale pipe sections, small samples were cut from the pipe, grooved by machining, gouging, or pressing, and tested under tension. Failure loads were measured as a function of groove depth and a simple model was sought to explain the results.

2. EXPERIMENTAL METHODS

In the present work, high strength linepipe steel CSA Z245.2 - M1979 Grade 386 (or equivalently high test linepipe steel API 5LX Grade 52) was used to prepare two types of specimens of nominal dimensions:

(a) 5.6 mm x 25.4 mm x 254.0 mm (thickness, width and span, respectively) called narrow specimens, Figure 1;

(b) 5.6 mm x 101.6 mm x 254.0 mm, called wide specimens, Figure 2. Specimens were cut with long axes parallel to the hoop direction, and flattened by pressing. U-O pipe was used, hence the rolling direction is parallel to the tensile axes.

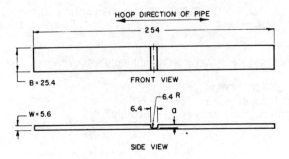

Figure 1 - *Narrow Specimen Geometry; Dimensions in mm*

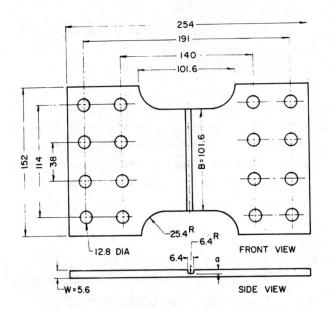

Figure: 2 - *Wide Specimen Geometry; Dimensions in mm*

Grooves were made by machining (to a depth up to 85 percent of plate thick
ness W), cold pressing (up to 55 percent of W), or gouging with a pendulum
scraper (up to 42 percent of W). Grooves were made in wide specimens by
machining only. Specimens were clamped in grips in an MTS servohydraulic
universal testing frame, and pulled at a constant displacement rate to
failure‾at room temperature.

3. EXPERIMENTAL RESULTS

To characterize the tensile properties of the steel, the constitutive equation $\sigma = k\varepsilon^n$ was fitted to true stress σ - true strain ε data for unnotched specimens. The best fit was obtained with k = 690 MPa and n = 0.12.

Results from tests of grooved specimens were plotted as the maximum gross stress σ^{GM} (load divided by original cross-sectional area WB of un-grooved region) as a function of normalized groove depth a/W in Figures 3, 4 and 5. Curves through the data points are based on a theoretical model developed in the next section. Measurements of Knoop indentation hardness KHN were made according to ASTM standard [1] on longitudinal cross-sections through samples with pressed grooves and gouges, and are reported in Figures 6, 7 and 8. The curve shown in Figure 6 is theoretically predicted, as explained in the next section.

4. DISCUSSION

"Plastic collapse" denotes a situation in which gross deformation of a body becomes possible. Collapse loads are most often calculated using a rigid-plastic flow model, with the flow stress taken as the material's yield stress σ^Y, or as $(\sigma^Y + \sigma^{UTS})/2$ (where σ^{UTS} is the ultimate tensile stress) to allow for work hardening [2,3]. In the present work, we are concerned with the maximum load-bearing capacity of the damaged plate, corresponding to ultimate plastic collapse analogous to the point of maximum load in a tensile test (i.e., at UTS) rather than to the onset of gross yield point in a tensile test. This allows the full work-hardening capacity of the material to be utilized.

The simplest possible model for the collapse stress, σ^{GM}, of a grooved plate is one in which slip planes form at 45° to the tensile axis at the reduced section, and hence, the failure load is simply reduced in proportion to the loss in cross-section area. The predictions of this model are plotted as σ^{GM} (UT), where UT denotes uniaxial tension, in Figures 3 to 5, being simply a straight line between σ^{UTS} for the ungrooved plate (a/W = 0) and zero at a/W = 1.0.

The most obvious feature of the results is that for all grooves investigated, σ^{GM} is significantly larger than σ^{GM} in uniaxial tension. This can be explained as a result of work hardening, as represented by the constitutive equation, and of constraint, because the material in the grooves deforms essentially under plane strain, whereas ungrooved plate deforms in uniaxial tension (provided the distance between grips is significantly larger, by at least a factor of 2 or 3, than the specimen width B).

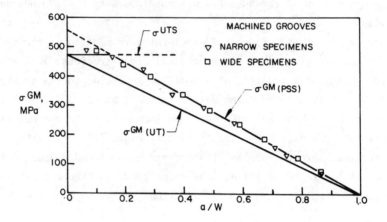

Figure 3 - *Dependence of* σ^{GM} *on Thickness Reduction in Specimens with Machined Notches*

FAILURE OUTSIDE GROOVE

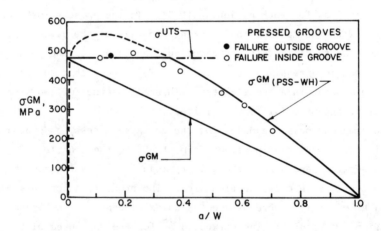

Figure 4 - *Dependence of* σ^{GM} *on Thickness Reduction in Specimens with Pressed Notches*

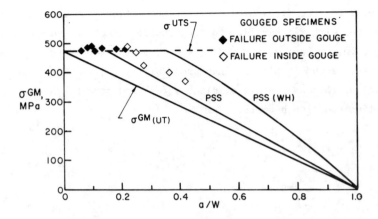

Figure 5 - Dependence of σ^{GM} on Thickness Reduction in Specimens with Scraped Grooves

Consider first the results of Figure 3, for a machined groove. In this case, the material under the groove is initially undeformed and has the same yield stress as the rest of the specimen. Therefore, yielding occurs first at the reduced section, and the unyielded plate on both sides of the groove constrains the deformation of the material at the grooved section to be in a state of plane strain (strain being essentially zero parallel to the direction of the groove). Then σ^{GM} can be found by using the constitutive equation, imposing plain strain deformation, and finding the stress at maximum load.

The appropriate constitutive equation must involve the three principle stresses $(\sigma_1, \sigma_2, \sigma_3)$ and strains $(\varepsilon_1, \varepsilon_2, \varepsilon_3)$ where σ_i and ε_i, i = 1,2,3 are true stresses and strains. This can be done by defining "effective" stresses and strains $\bar{\sigma}$ and $\bar{\varepsilon}$ given by [4]

$$\bar{\sigma} = \frac{\sqrt{2}}{2} \left[(\sigma_1-\sigma_2)^2+(\sigma_2-\sigma_3)^2+(\sigma_3-\sigma_1)^2 \right]^{1/2} , \tag{1}$$

and

$$\bar{\varepsilon} = \frac{\sqrt{2}}{3} \left[(\varepsilon_1-\varepsilon_2)^2+(\varepsilon_2-\varepsilon_3)^2+(\varepsilon_3-\varepsilon_1)^2 \right]^{1/2} , \tag{2}$$

with these definitions, and using $\bar{\sigma} = k\bar{\varepsilon}^n$, the values of k and n can be found from uniaxial tensile tests for which $\sigma_1 = \sigma$, $\sigma_2 = \sigma_3 = 0$, $\varepsilon_1 = \varepsilon$, $\varepsilon_2 = \varepsilon_3 = -\varepsilon_1/2$, and so $\sigma = k\varepsilon^n$. As reported above, k = 690 MPa and n = 0.12 for the steel used. For the conditions depicted in Figure 3, choosing axes with x_1, parallel to the tensile direction and x_3 parallel to the groove, we have $\sigma_2 = 0$ and $\sigma_3 = \sigma_2/2$ by the Levy-Mises equations and $\varepsilon_2 = -\varepsilon_1$, assuming plastic incompressibility [5].

Labelling this as a state of plane stress and strain PSS, the above constitutive equation becomes:

$$\sigma_1 = k'\varepsilon_1{}^n , \tag{3}$$

where

$$k' = \left(\frac{2}{\sqrt{3}}\right)^{1+n} k . \tag{4}$$

Values of k'/k are insensitive to changes in n over the range 0.1 to 0.2, and for n = 0.12 we find k'/k = 1.175. Solving for the value of σ_1, at which $d\sigma_1/de_1 = 0$ (where e_1, is the nominal strain) using equation (3), it was found that the maximum load is reached at the strain $\varepsilon_1^* = n$, which is the same result as obtained for uniaxial tension. In these circumstances, the maximum nominal stresses under the constraints appropriate for PSS and UT are $k'(n/e)^n$ and $k(n/e)^n$ respectively, where e is the base of natural logarithms. The ratio of these stresses is k'/k = 1.175 for n = 0.12, and so values of σ^{GM} (PSS) are predicted to be 17.5 percent higher than σ^{GM} (UT) It can be seen in Figure 3 that measurements are closely in accord with this deduction for a/W above 0.15. For a/W below 0.15, failure stress in the groove is predicted to be higher than failure stress of the ungrooved material, and so of course the collapse stress under these conditions is σ^{GM} (UT).

Note also in Figure 3 that data for wide and narrow specimens are in close agreement. This supports the argument that the geometry in both cases (groove length to groove width ratio of 4 and 16 for narrow and wide specimens respectively) is adequate to maintain PSS constraints on the groove cross section.

Consider next the data for the specimens with pressed grooves, Figure 4 In this case, the material in the groove is uniformly hardened by deformation during the pressing operation; deformation is done under plane strain conditions, with $\varepsilon_3 = 0$. Hence, the flow stress can be predicted from the constitutive equation using the depth of the pressed groove to obtain the as-pressed strain from $\varepsilon_2 = \ln(1-a/W) = -\varepsilon_1$. Making the reasonable assumption that the hardness H is proportional to the prestressing so that $H = C\varepsilon^{-n}$ where C is a constant, the curve through the points in Figure 6 may be derived using C = 282 KHN. The fit to experimental data is seen to be reasonable. Also, for deformation in uniaxial tension of ungrooved samples, where $\varepsilon_2 = \varepsilon_3 = -\varepsilon_1/2$, this method predicts hardnesses of 162 KHN and 208 KHN at $\varepsilon_1 = 0.01$ and 0.12 (strain at UTS) respectively. These are in reasonable agreement with the measured hardness values at yield and UTS in Figure 6.

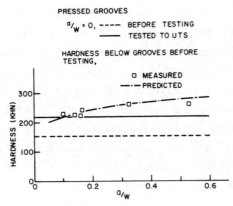

Figure 6 - Average Hardness of Material Below Pressed Notched of Different Depths

After the pressing operation, specimens were tested in tension. As in the case of machined notches, deformation will be concentrated in the groove provided that the load-bearing capacity of the grooved region is less than that of the ungrooved region. In this case, deformation will continue under PSS conditions as for the machined groove, but in this case the material at the groove is initially hardened. Because both pressing and testing are done under plane strain, the deformation history of the grooved region is in fact the same as that of a specimen deformed entirely under PSS conditions. If the pressed groove is deeper than the value of a/W, corresponding to the strain at which maximum load is reached during PSS deformation of an ungrooved sample, then maximum load is reached immediately upon tensile loading of the specimen containing the pressed groove. It follows that:

$$\sigma^{GM} = k' \; [-\ln \; (1-a/W)]^{n} (1-a/W) \; . \tag{5}$$

The curve given by equation (5) is shown in Figure 4 labelled PSS-WH (for plane stress and strain with work hardened material), and it may be seen that agreement with experiment is very good. At values of a/W < 0.35, the maximum load is limited by the uniaxial tensile strength of the ungrooved region, and failure can occur outside or inside the groove.

Finally, for gouged specimens, both loss of section and work hardening affect the maximum load-bearing capacity. It is known from the hardness measurements shown in Figures 7 and 8 that hardening occurs through the entire section below a gouge. Indeed, the average hardness below the gouge (excluding measurements in the first x/W = 0.04) in samples for which $0.2 \leq a/W \leq 0.33$ is about 240 KHN, which is only slightly less than for specimens with grooves pressed to depths in this range (Figure 6). There is a severely hardened region very near the surface of a gouge, with hardnesses reaching 500 KHN. This corresponds to a surface layer that is transformed metallurgically as well as mechanically by the severe shocks experienced at the surface. However, the transformed layer is very shallow extending no more than 0.2 mm below the surface.

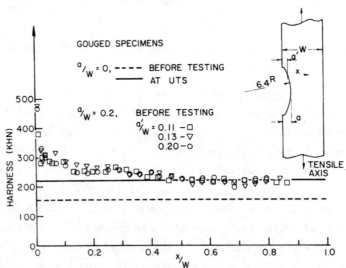

Figure 7 - Hardness Measurements for Gouged Specimens as a Function of Distance below the Gouge Surface Normalized by Plate Thickness along Three Different Traverses of the Remaining Ligament. Also shown are hardness levels of ungrooved material, untested and strained to UTS.

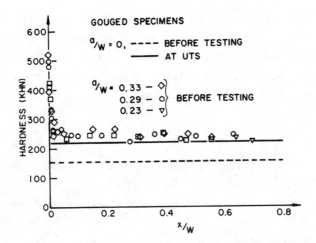

Figure 8 - Hardness Requirements below Gouged Notches of Three Different Gouge Depths

The predicted collapse loads for the PSS and PSS-WH conditions shown in Figures 3 and 4 are reproduced in Figure 5 along with measured values. It is evident that results for gouged specimens lie between those for specimens with machined and pressed grooved, as would be expected from the above discussion.

Perhaps surprisingly, these results have demonstrated an increase in the collapse stress in the grooved region for machined, pressed, and gouged grooves compared with the UTS in a tensile test. However, it must be remembered in using these results to deduce failure loads of damaged pipelines that the constraints in this case are those of plane strain for undamaged pipe, rather than uniaxial tension. Hence, the appropriate collapse stress for undamaged pipe is about 17 percent higher than the UTS.

5. CONCLUSIONS

(1) The effect of a machined groove is to increase the collapse stress on the grooved region by about 17 percent above the σ^{GM} (UT) of ungrooved material, for grooves deeper than a/W = 0.15.

(2) The collapse stress on the section containing a pressed groove is higher by at least 50 percent than the UTS, for grooves deeper than a/W = 0.35.

(3) The collapse stress for gouged specimens lies between the values for specimens containing grooves machined and pressed to the same depth.

(4) These observations are consistent with an explanation based on work hardening, using a simple constitutive equation, and taking into account the constraints of the deformation.

REFERENCES

1.   "Microhardness of Materials", *ASTM E384-63*, 1979.
2.   WILLOUGHBY, A.A., *A Survey of Plastic Collapse Solutions Used in the Failure Assessment of Part Wall Defects*, The Welding Institute, Cambridge, 1982.
3.   CHELL, G.G., "Elastic-Plastic Fracture Mechanics", *Developments in Fracture Mechanics - 1*, Applied Science Publishers Ltd., London, 1979, pp. 67-105.
4.   NADAI, A., *Theory of Flow and Fracture of Solids*, McGraw Hill Book Company, New York, Vol. 1, 1950.
5.   DIETER, G.E., *Mechanical Metallurgy*, Second Edition, McGraw-Hill, New York, 1976.

MODELLING PROBLEMS IN CRACK TIP MECHANICS   *Waterloo, Ontario, Canada*
CFC10, University of Waterloo   *August 24-26, 1983*

TRANSIENT STRAIN FIELDS OF DUCTILE FRACTURE OF A
CENTRALLY CRACKED ALUMINUM SHEET

R. WILLIAMS and F.P. CHIANG
Department of Mechanical Engineering
State University of New York at Stony Brook
Stony Brook, Long Island, New York, U.S.A.

1. INTRODUCTION

In the study of fracture mechanics many analytical difficulties are
encountered in the modelling of a ductile material's fracture characteristics
whose geometrical configuration closely resembles that of plane stress.
Theoretical works such as the one presented by Rice [1], provide solutions
for small scale yielding about the crack tip subject to several types of
loading conditions in which most of the analysis is made with regard to
the geometrical consideration of plane strain. Theories dealing with crack
propagation [2], are usually examined by energy methods which may adequately
describe brittle fracture, but has limited application to ductile, polycrystalline engineering alloys. To avoid many of these theoretical compli-
cations, the moire method can be readily applied to many types of fracture
specimens as a means of observing the large deformations associated with
the fracture of ductile engineering alloys. Liu and Ke [3] have provided
a comprehensive treatise on the application of the in-plane moire method
to examine the fracture characteristics of a variety of specimen configurations and loading parameters.

In this paper both the in-plane moire and shadow moire methods are
used to obtain whole-field contours for the displacement components u, v
and w. These techniques are applied to a centrally cracked aluminum sheet,
which upon loading will undergo large scale yielding prior to and during
fracture. By selection of the appropriate grating pitch, moire can easily
accommodate displacement measurements of the large deformations encountered.
Recording the resulting moire fringe patterns with a high speed camera,
fringe data can be extracted for analysis at various times during the
deformation and fracture history.

2. EXPERIMENTAL SET-UP

The specimen configuration used in this work is a tensile specimen
fabricated from 3003 H-14 aluminum sheel 3.18 mm (1/8") thick. An elec-
trically discharged slit located in the centre of the specimen was chosen
as a stress concentrator, shown in Figure 1. Three identical specimens
are used to individually record the in-plane displacement components u
and v, and the out-of-plane displacement component w. The need to individually record each of the in-plane displacement components is due to the
inability of the rotating prism high speed camera used in this experiment,
to resolve the continuously deforming specimen grating. This experimental
difficulty prohibits the use of a cross grating to simultaneously record
both in-plane displacement fields on a single specimen and then separating
them with the use of the optical spatial filtering arrangement [4]. With
the use of two specimens to determine the in-plane displacement components
line gratings each having a pitch of 4 lines/mm (100 lines/inch) were
printed with the use of Kodak (KPR) photoresist on the surface of the
specimen. The specimen used to record the u-field isothetics or the moire
fringe pattern which represents the x-direction displacement field has
the principal direction of the printed line grating coincident with the
x-axis. Similarly the specimen used to record the v-field isothetics has
its grating orientation such that the principal direction is coincident
with the y-axis. The third specimen used to measure the out-of-plane
displacement requires no special preparation except, blasting with a fine
glass bead to produce a dull nonreflective surface which yields enhanced
contrast to the shadow moire fringes.

The specimens were then subject to a uniaxial tensile load at a prescribed strain rate according to the experimental set-up shown in Figure 2.
The loading parameters in all these tests were displacement of the hydraulic
cylinder versus time in the form of a linear ramp. This results in a constant cylinder velocity 4.5 cm/sec (1.77 in/sec) from the time loading
begins until fracture. The framing rate controlled by the Fastax Goose
Control was kept constant throughout the event at 1000 frames per second.
The pulsed square wave generator triggers a neon strobe inside the camera
which produces timing marks on the border of the film by which films containing different fields of measurement can be referenced. An x-y recorder
is used to plot the measured load versus time for the entire event.

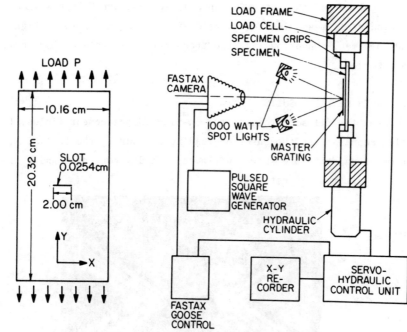

Figure 1
Specimen Configuration Used

Figure 2
Experimental Arrangement for Recording
In-Plane Displacement Fields

To obtain the in-plane displacement fields the master grating from which the specimen grating is printed, is directly superimposed over the specimen and adjusted until no rotational mismatch is observed. The master grating is then clamped to the upper specimen grip which remains rigid during the test. During loading the deforming specimen grating interferes with stationary master grating producing a transient moire fringe pattern which was illuminated by two 1000 watt spot lights and recorded by the high speed camera. This procedure was repeated for both in-plane specimens keeping all loading and photographic recording parameters identical except for the orientation of the specimen grating, master grating and lights such that both the u and v displacement fields were obtained. The third specimen was again subjected to the same experimental parameters with only a glass master grating fixed parallel to the surface of the specimen and a single light source. Prior to loading the experimental shadow moire arrangement [5] was calibrated using a wedge of known dimensions and identical

orientation to the test specimen. This procedure immediately determines
the magnitude of the out-of-plane displacement per fringe without resorting
to accurately measuring the large distances between camera, light source
and specimen. Upon loading, the transient out-of-plane displacement field
is recorded.

3. EXPERIMENTAL RESULTS

From the films containing the in-plane displacement fields, frames 175
milliseconds after the onset of loading and during the initial stages of
crack propagation are shown in Figures 3 and 4 for the u and v field iso-
thetic patterns respectively.

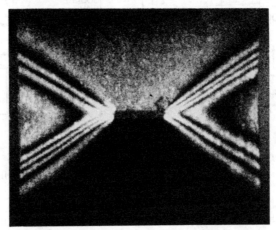

Figure 3 - u-Field, Dynamic In-Plane Moire Pattern

Figure 4 - v-Field, Dynamic In-Plane Moire Pattern

The moire fringe patterns represent contours of constant in-plane displacement component whose magnitude is governed by the following equations
[6]:

$$u = Np , \quad \text{and} \quad v = Mp ,$$

(1),(2)

where p is grating pitch, which is identical for both fringe patterns and
N and M are the fringe orders for the u and v field patterns respectively.
Since the grating pitch is a constant value, **the derivatives of displacement**
with respect to position needed to formulate the components of strain ε_{xx},
ε_{yy} and ε_{xy} rely on determining the four partial derivatives $\partial N/\partial x$, $\partial M/\partial x$,
$\partial N/\partial y$ and $\partial M/\partial y$. Since the deformation endured by these specimens is large,
strain values were calculated using the Lagrangian description for finite
strain according to the following equations:

$$\varepsilon_{xx} = \frac{\partial u}{\partial x} + \frac{1}{2}\left[\left(\frac{\partial u}{\partial x}\right)^2 + \left(\frac{\partial v}{\partial x}\right)^2\right] ,$$

(3)

$$\varepsilon_{yy} = \frac{\partial v}{\partial y} + \frac{1}{2}\left[\left(\frac{\partial v}{\partial y}\right)^2 + \left(\frac{\partial u}{\partial y}\right)^2\right] ,$$

(4)

$$\varepsilon_{xy} = \frac{1}{2}\left[\frac{\partial v}{\partial x} + \frac{\partial u}{\partial y} + \frac{\partial u}{\partial x}\frac{\partial u}{\partial y} + \frac{\partial v}{\partial x}\frac{\partial v}{\partial y}\right] .$$

(5)

To determine the partial derivatives of fringe order with respect to
position such that the previous strain displacement relations can be evaluated, integer fringe orders are assigned to both fields of measurement
according to the rules set forth in [6,7]. With each fringe pattern
assigned its appropriate fringe orders the origin of a cartesian coordinate
system is chosen as the centre of the stress concentrator. From this
origin both u and v field fringe orders are interpolated using the smoothed
spline function [8] as a function of y for specific values of x ranging
from just outside the crack tip to the boundary of specimen for both the
left and right sides of the specimen. As an example, consider interpolating
the v-field fringe order M as a function of y at a specific x-coordinate
using the smoothed spline function. The smoothed spline is a piecewise-
interpolation polynomial of order 3 with continuous first and second
derivatives defined over the interval of the abscissae (y_0, y_n). The
smoothing function M(y) interpolates fringe order as the ordinate with an
individual cubic polynomial along each sub-interval (y_i, y_{i+1}) of the form

$$M(y) = a_i + b_i(y-y_i) + c_i(y-y_i)^2 + d_i(y-y_i)^3 , \qquad (6)$$

where

$$y_i \leq y < y_{i+1} \qquad \text{and} \qquad 0 \leq i \leq n-1 .$$

The coefficients a_i, b_i, c_i and d_i are determined such that the integral:

$$\int_{y_0}^{y_n} \frac{d^2M(y)^2}{dy^2} \, dy , \qquad (7)$$

is minimized with the condition that:

$$\sum_{i=0}^{n} \left[\frac{M(y_i)-M_i}{\delta M_i} \right]^2 \leq S . \qquad (8)$$

In terms of a mechanical spline or from beam theory equation (7) states that the spline assumes the shape which minimizes its potential energy. The condition set forth in equation (8) allows the spline to deviate from the assigned fringe order M_i or knots controlled by the value of the smoothing index δM_i and the redundant smoothing parameter S which determine the extent of smoothing. For this analysis the coefficients of the cubic polynomial were obtained by the solution and algorithm formulated by Reinsch [8], with the smoothing index δM_i taken as the error in reading the position of the fringe order (.005) and the value of S equal to the number of data points to be interpolated along each spline. Upon determination of the four spline coefficients the derivative of fringe order with respect to position is easily found to be:

$$\frac{d\ M(y)}{dy} = b_i + 2c_i(y-y_i) + 3d_i(y-y_i)^2 . \qquad (9)$$

The application of the spline function for determining good quality derivatives from experimental data can be found in [9].

Once the fringe orders N and M have been interpolated with respect to y for several x-positions common to both specimens, fringe order and its derivatives with respect to y can be calculated along the specified intervals. Choosing y-positions within these intervals again common to both specimens the fractional fringe orders at some constant y-position for each half of the specimen can be interpolated as a function x. The result is a rectangular grid of identical dimensions for both the u and v field

specimens where fringe order and its derivatives with respect to x and y
are known at the nodal points from which data was originally taken. If
the two grids are superimposed with the assumption that two identical
specimens deform identically keeping all experimental loading parameters
constant, the four derivatives necessary to formulate strain according to
equations (3,4,5) are known at the nodal points of the grid. Upon formula-
tion of the strain components ε_{xx}, ε_{yy} and ε_{xy} at the coincident nodal
points the coordinates for specific values of strain were inversely inter-
polated using linear interpolation for each interval of strain as a constant
x-position of the grid. Following this procedure for each successive
x-position, contours of constant strain were constructed for all three
in-plane strain components as shown in Figure 5.

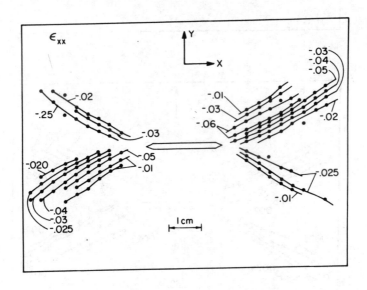

(a)

*Figure 5 - Contour of Constant Strain 175 milliseconds after the Onset
of Loading*

    (a) strain contours of ε_{xx},
    (b) strain contours of ε_{yy},
    (c) strain contours of ε_{xy}.

(continued)

(continued)

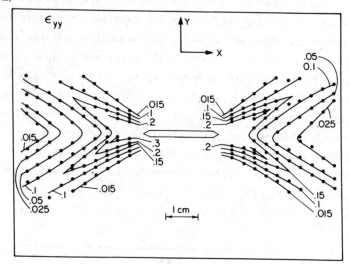

(b)

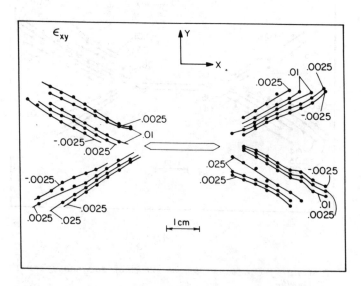

(c)

Figure 5 - *Contour of Constant Strain 175 milliseconds after the Onset*
of Loading

  (a) *strain contours of* ε_{xx},
  (b) *strain contours of* ε_{yy},
  (c) *strain contours of* ε_{xy}.

From the third film containing the transient shadow moire fringe
patterns, the frame 175 milliseconds after the onset of loading is shown
in Figure 6. Fringes are ordered with integer values starting with a
value of one for the fringe furthest from the crack tip and successively
increasing by one for each fringe encountered from the fringe of order
one to the crack tip. For the fringe pattern shown in Figure 6, the
greatest observable fringe value is 3. Since each shadow moire fringe
represents a contour of constant out of plane displacement and if the
following expression for the out-of-plane strain component ε_{zz} is used:

$$\varepsilon_{zz} = \frac{\partial w}{\partial z} \simeq \frac{\Delta w}{\Delta z} \ . \tag{10}$$

The shadow moire fringe then represents a contour of constant out-of-plane
strain whose magnitude is approximately equal to:

$$\varepsilon_{zz} \simeq \frac{-2\delta}{\tau} \ , \tag{11}$$

where δ is the magnitude of the out-of-plane displacement for a specific
fringe order determined by the calibration wedge and τ is the thickness
of the specimen. Table 1 shows the magnitude of the out-of-plane displacement and the corresponding strains for the fringe pattern of Figure 6.

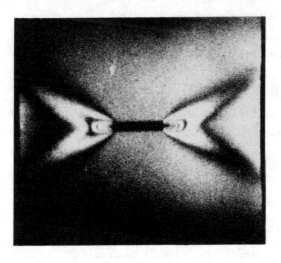

Figure 6 - w-Field, Dynamic Shadow Moire Pattern

Table 1 - *Strain Values (ε_{zz}) Corresponding to the Fringe Patterns in Figure 6*

Fringe Order	Out-of-Plane Displacement (δ)	Strain ε_{zz}
1	.0076 cm	-.0478
2	.033 cm	-.2075
3	.061 cm	-.3836

4. CONCLUSIONS AND DISCUSSION

We have demonstrated the capability of using both in-plane and shadow moire techniques to map the transient strain fields surrounding a running ductile crack. Due to experimental difficulties we were forced to separately obtain each displacement field u, v and w on individual specimens. However, all three displacement fields can be determined on a single specimen with the use of more refined experimental techniques which have and are currently being carried out on static test specimens with future plans for extending them to dynamic analysis. While this paper concentrate on the moire technique applied to ductile fracture, this analysis can be extended to experimental data at several time intervals during fracture of the specimen. In addition to the results presented here, other fractur parameters such as crack extension, crack propagation velocity, crack opening displacement and evaluation of the J-integral can be obtained from this experimental data.

ACKNOWLEDGEMENTS

The work reported here was supported in part by the Office of Naval Research Contract No. N0001482K0566.

REFERENCES

1. RICE, J.R., "Mathematical Analysis in the Mechanics of Fracture", *Fracture*, edited by H. Liebowitz, Vol. 2, 1968, pp. 191-311.
2. ERDOGAN, F., "Crack Propagation Theories", *Fracture*, edited by H. Liebowitz, Vol. 2, 1968, pp. 497-590.
3. LIU, H.W. and KE, J.S., "Moire Method", *Experimental Techniques in Fracture Mechanics*, edited by A.S. Kobayashi, Vol. 2, Society of Experimental Stress Analysis Monograph No. 2, 1975.
4. CHIANG, F.P., "Techniques of Optical Spatial Filtering Applied to the Processing of Moire Fringe Pattern", *Experimental Mechanics*, Vol. 9, No. 11, 1969, pp. 523-526.

5. THEOCARIS, P.S., *Moire Fringes in Strain Analysis*, Pergamon Press Inc.
    1969.

6. CHIANG, F.P., "Moire Method of Strain Analysis", *Experimental Mechanics*,
    Vol. 19, No. 8, 1979, pp. 290-308.

7. DURELLI, A.J. and PARKS, V.J., *Moire Analysis of Strain*, Prentice Hall,
    Inc., 1970.

8. REINSCH, C.H., "Smoothing by Spline Functions", *Numerische Mathematik*,
    Vol. 10, 1967, pp. 177-183.

9. BERGHAUS, D.G. and CANNON, J.P., "Obtaining Derivatives from Experi-
    mental Data Using Smoothed-Spline Functions", *Experimental Mechanics*,
    Vol. 13, No. 1, 1973, pp. 38-42.

Author Index